A Text Book Of

MEASUREMENT AND AUTOMATION

(Subject Code : MEC606)

Semester - VI

THIRD YEAR DIPLOMA IN MECHANICAL ENGINEERING GROUP

As Per SBTE's New Revised Syllabus

VINOD THOMBRE-PATIL

B.E. (Mech.) M. Tech. (Energy Technology)
Former Vice Principal and HOD - Mechanical,
S. H. Jondhale Polytechnic,
Dombivli, Dist. Thane, Maharashtra.

PRASHANT K. AMBADEKAR

M.E., LMISTE, AMIE
Assistant Professor in Mechanical Department,
South Indian Education Society,
Graduate School of Technology
Nerul, Navi Mumbai.

APEKSHA THOMBRE-PATIL

B.E., M.E. (Electronics and Telecommunication),
Assistant Professor,
G.V. Acharya College of Engineering and Technology,
Shelu, Dist. Thane, Maharashtra.

N4739

Measurement and Automation   ISBN 978-93-89944-16-7

First Edition   :   February 2020
©   :   Authors

Published By :
NIRALI PRAKASHAN
Abhyudaya Pragati, 1312, Shivaji Nagar
Off J.M. Road, PUNE – 411005
Tel - (020) 25512336/37/39, Fax - (020) 25511379
Email : niralipune@pragationline.com

➢ **DISTRIBUTION CENTRES**

PUNE

Nirali Prakashan : 119, Budhwar Peth, Jogeshwari Mandir Lane, Pune 411002, Maharashtra
Tel : (020) 2445 2044
Email : bookorder@pragationline.com, niralilocal@pragationline.com

Nirali Prakashan : S. No. 28/27, Dhyari, Near Pari Company, Pune 411041
Tel : (020) 24690204 Fax : (020) 24690316
Email : dhyari@pragationline.com, bookorder@pragationline.com

MUMBAI

Nirali Prakashan : 385, S.V.P. Road, Rasdhara Co-op. Hsg. Society Ltd.,
Girgaum, Mumbai 400004, Maharashtra
Tel : (022) 2385 6339 / 2386 9976, Fax : (022) 2386 9976
Email : niralimumbai@pragationline.com

➢ **DISTRIBUTION BRANCHES**

JALGAON

Nirali Prakashan : 34, V. V. Golani Market, Navi Peth, Jalgaon 425001,
Maharashtra, Tel : (0257) 222 0395, Mob : 94234 91860
Email : niralijalgoan@pragationline.com

KOLHAPUR

Nirali Prakashan : New Mahadvar Road, Kedar Plaza, 1st Floor Opp. IDBI Bank
Kolhapur 416 012, Maharashtra. Mob : 9850046155
Email : niralikolhapur@pragationline.com

NAGPUR

Nirali Prakashan : Above Maratha Mandir, Shop No. 3, First Floor,
Rani Jhanshi Square, Sitabuldi, Nagpur 440012, Maharashtra
Tel : (0712) 254 7129; Email : niralinagpur@pragationline.com

DELHI

Nirali Prakashan : 4593/15, Basement, Agarwal Lane, Ansari Road, Daryaganj
Near Times of India Building, New Delhi 110002 Mob : 08505972553
Email : niralidelhi@pragationline.com

BANGALURU

Nirali Prakashan : Maitri Ground Floor, Jaya Apartments, No. 99, 6th Cross, 6th Main,
Malleswaram, Bangaluru 560 003, Karnataka
Mob : +91 9449043034

Other Branches : Hyderabad, Chennai

niralipune@pragationline.com | www.pragationline.com

Also find us on   www.facebook.com/niralibooks

Preface ...

We are glad to present the book entitled **"Measurement and Automation"** for **IIIrd Year (Sixth Semester)** Diploma in Mechanical Engineering as per the syllabus prescribed by SBTE.

We have observed the students facing extreme difficulties in understanding the basic principles and fundamental concepts without adequate solved problems along with the text. To meet this basic requirement of students, sincere efforts have been made to present the subject matter with frequent use of figures and lots of numerical examples.

We are thankful to Mr. Dineshbhai Furia, Mr. Pradeepbhai Furia, Mr. Jigneshbhai Furia, Mr. Shashikant Patel, Mr. Akbar Shaikh, Mrs. Manasi Pingle, Mrs. Anjali Muley, Mr. Ravindra Walodare of Nirali Prakashan along with their staff members for bringing this publication timely to the Market.

We must acknowledge special gratitude to Mr. Raosaheb Patil Danve (Minister of State, Govt. of India), Er. Rajesh Tope (MLA and President of Matsyodari Shikshan Sanstha, Maharashtra), Er. Sagar Jondhale (Director, Jondhale Educational Group), Dr. Manoj Shete (President, College of Engg and Tech., Kasara), Dr. Ram Reddy, (Principal, LRTCOE, Mumbai), Dr. R. G. Tated, Prof. Shinde N. N. (HOD - Energy Technology, DOT, SUK), Prof. Sharad Pawar (Principal, Fabtech Polytechnic), Dr. Nehte (Principal, Father Agnel), Dr. Pable M. J. (G. P. Bandra), Prof. R. B. Sholapurkar, (Principal, Virar Polytechnic), Prof. Taskar (Principal, G.P., Jintur), Prof. Kantute (Principal, SBM, Mumbai), Prof. Bavaskar (Principal, Muchala Polytechnic, Thane), Prof. Tandon (Principal, BVIT, Mumbai), Prof. Mange Amar (Principal, Pillai Polytechnic, Rasayani), Prof. Tambe (Principal, Shantiniketan Polytechnic, Panvel), Prof. S. N. Mahajan (H.O.D., V.B.V.P., Vasai), Prof. Anoop Kumar and Prof. Sawant (H.O.D., S.S.J.P., Ambernath), Prof. Mrs. Nehal Muchala (HOD, Thakur Polytechnic), Prof. Sable (HOD, Tasgaonkar Polytechnic, Karjat), Prof. Kamble (HOD, Navjeevan Polytechnic), Prof. S. S. Bhadane (HOD, SSJ Polytechnic, Asangaon), Prof. A. S. Dayma (HOD, SSJCOE, Asangaon), Prof. K. A. Jawalkar (HOD, SSJ Polytechnic, Asangaon), Prof. M. G. Bagale (HOD, SPM Polytechnic, Kunthe, Dist. Solapur), Prof. Killedar (G.P., Solapur), Prof. Lonkar, Prof. Salunkhe, Prof. Sanghvikar, Prof. P. P. Chaudhari, Prof. Shele, Prof. Pandit, (G.P., Thane), Mr. Anish Gunjal (M. Tech., I.I.T., Mumbai) and many more, who have extended their co-operation and suggestions time to time.

We specially thank our students for motivating and encouraging us from time to time.

Any errors or suggestions for the improvement of this book brought to notice will be thankfully acknowledged and incorporated in the next edition.

Write at: vinodthombrepatil.author@outlook.com

prashantambadekar@gmail.com

Vinod Thombre-Patil

Prashant Ambadekar

Apeksha Thombre-Patil

■■■

Syllabus ...

1. Introduction to Measurement   (Hours 08)

- Significance of Measurement, Types of Measurement, Classification of Instruments.
- Static Terms and Characteristics - Range and Span, Accuracy and Precision, Reliability, Calibration, Hysteresis and Dead Zone, Drift, Sensitivity, Threshold and Resolution, Repeatability and Reproducibility, Linearity.
- Dynamic Characteristics - Speed of Response, Fidelity and Dynamic errors, Over shoot.
- Measurement of Error - Classification of Errors, Environmental Errors, Signal Transmission Errors, Observation Errors, Operational Errors.

2. Displacement Measurement   (Hours 08)

- Capacitive Transducer, Potentiometer, LVDT, RVDT, Specification, Selection and Application of Displacement Transducer.

Temperature Measurements

- Non-electrical Methods - Bimetal and Liquid in Glass Thermometer, Pressure Thermometer.
- Electrical Methods - RTD, Platinum Resistance Thermometer, Thermistor, Thermo Electric Methods - Elements of Thermo couple, Law of Intermediate Temperature, Law of Intermediate Metals, Thermo EMF Measurement.
- Quartz Thermo Meter.
- Pyrometers - Radiation and Optical.

3. Flow Measurements   (Hours 06)

- Variable Head Flow Meters - Venturi, Flow nozzle, Orifice plate, Pitot Tube.
- Variable Area Meter - Rota Meter.
- Variable Velocity Meter - Anemometer.
- Special Flow Meter - Hot Wire Anemometer, Electromagnetic Flow Meter, Ultrasonic Flow Meter.

Note: Simple Numerical on above Topics.

4. Miscellaneous Measurement   (Hours 08)

- Force and Shaft Power Measurement - Tool Dynamometer (Mechanical Type), Eddy Current Dynamometer, Strain Gauge Transmission Dynamometer.
- Speed Measurement - Eddy Current Generation Type Tachometer, Incremental and Absolute Type, Mechanical Tachometers, Revolution Counter and Timer, Slipping Clutch Tachometer, Electrical Tachometers, Stroboscope.
- Strain Measurement - Stress-strain Relation, Types of Strain Gauges, Strain Gauge Materials, Selection and Installation of Strain Gauges, Load Cells, Rosettes.

Note: Simple Numerical on above Topics.

5. Automation   (Hours 04)

- Basic Elements of Automated System, Advanced Automation Functions, Levels of Automation.
- Flexible Manufacturing System - Introduction, Scope and Benefits, Types.
- Major Elements of FMS, FMS equipment, FMS Application.
- Introduction to CIM.

6. Robotics   (Hours 08)

- Definition, Robot Anatomy, Classification of Robots.
- Sensors - Contact and Non-contact, Touch, Tactile, Range and Proximity Sensor.
- End Effectors, Types of End Effectors, Robot Programming Languages, Robot drives, Applications of Robots, One Specific Application of Industrial Robot, Material Handling, Automated Guided Vehicle System.

■■■

Contents ...

■■■

Chapter 1

INTRODUCTION TO MEASUREMENT

1.1 MEASUREMENT

- **Measurement** is defined as, *"an act or result of comparison between a quantity (whose magnitude is unknown), with a similar quantity such as standard (whose magnitude is known)".*

- In other words, measurement is defined as, **"the process of determining the value of magnitude of an unknown quantity by comparing it with some standard or reference".**

- Since the two quantities, one unknown and one known are compared, the result is expressed in terms of a numerical value. Quantity to be measured, i.e. unknown quantity is called as 'Measurand' and known quantity is called as 'Reference or Standard'.

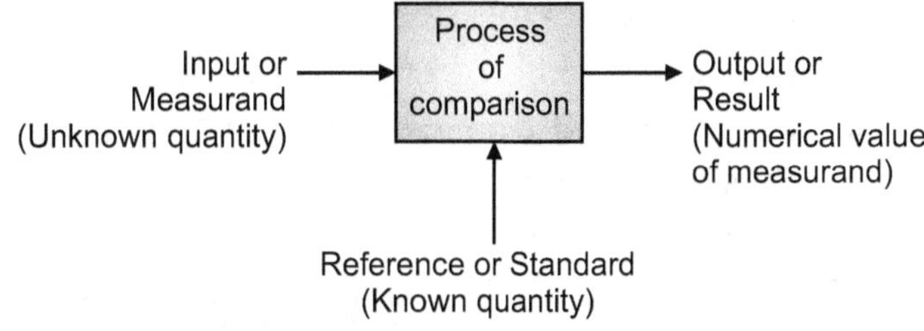

Fig. 1.1: Process of Measurement

1.2 BASIC REQUIREMENTS FOR MEANINGFUL RESULTS OF MEASUREMENT

1. The standard should be accurately defined and universally accepted.
2. The standard must be of same characteristic, as the measurand is.
3. The apparatus used and method adopted for the purpose of comparison must be provable.

1.3 SIGNIFICANCE OF MEASUREMENT

1. Measurement is the primary source of information on various aspects in design, production, processing etc.
2. The product quality and process efficiency are as a result of measurement and corrective action, if any.
3. Measurement plays a very significant role in process industries and power plants, which include control system, process instrumentation and data reduction.
4. Measurement covers various aspects such as detection, sensing, data manipulation, transmission, acquisition, control and analysis of data. The qualities involved in measurement may be physical, mechanical, electrical, optical etc.
5. In the field of engineering design and R & D programs, the measurement and its correct interpretation are the sources of great importance.
6. Measurement provides a standard unit for every kind of quantity being measured. Therefore, it becomes possible to express the measured quantity in terms of a numerical value or meaningful number.
7. The modern era of automation and automatic controls used in manufacturing system are fully dependent on the measurement and control of different parameters. Here, the act of measurement fulfills this requirement satisfactorily.
8. Due to the measurement system, the product is designed and manufactured with maximum efficiency and minimum cost.

9. Measurement system provides feedback information necessary for proper operation and maintenance of manufacturing system and process. This helps in achieving goals and objectives of engineering and technology.

10. Measurement forms an essential basis of commercial activities such as pricing, sale and purchase.

11. Measurement of various parameters and a change in their magnitude can serve as a diagnostic aid with respect to the safety and health of men and machines and can be used for forecasting malfunctioning and failure.

12. Measurement plays a significant role in achieving goals and objectives of technology, because of the feedback information supplied by following functions:

    (i) Design of instrument and process.

    (ii) Proper maintenance of instrument and process.

1.4 OBJECTIVES OF MEASUREMENT

1. To specify/measure certain quantity, e.g. Diameter of rod, Pressure of fluid etc.

2. To check validity of design.

3. To monitor and control the variables involved in process.

4. To study and analyze different elements of machine structure using different laws of physics, mechanics etc.

5. To test the quality of product.

1.5 METHODS OF MEASUREMENTS

- Methods of measurements can be classified in two ways as given below.

    1. Direct and Indirect comparison methods.

    2. Primary, secondary and tertiary measurements.

1.5.1 Direct and Indirect Comparison Methods

(a) Direct comparison method:

- In this method, the parameter or quantity to be measured is directly compared with either a primary or a secondary standard.

- Direct comparison is quite commonly used for measurements of length. However, for measurement of mass, the problem becomes much more complicated since it is just not possible for human beings to distinguish between wide margins of mass.

- This method is less accurate and less sensitive. Therefore, measured values are subjected to error.

(b) Indirect comparison method:

- In this method, the comparison is made with a reference or standard through the use of a calibrated system.

- These methods for measurements are used in those cases, where the desired parameter to be measured is difficult to be measured directly, but it has got some correlation with some other parameter, which can be easily measured.

- **For example:** The elimination of bacteria in milk is directly dependent upon its temperature. Thus, the bacteria elimination can be measured indirectly by measuring the temperature of the milk.

- In indirect measurements, an empirical relation is generally established between the measurement actually made and the results, which are desired.

- When the primary purpose of making a measurement is to determine quality of a product, then quality should be measured directly. However, in case, direct measurement is not possible, then, indirect method of measurement should be used.

- Measurement system used in engineering applications make use of **indirect methods** for measurement purposes.
- A measurement system consists of a transducing element, which converts the quantity to be measured with an analogous signal. This signal after being processed by some intermediate means is then fed to the end devices, which provide the measurement results.

1.5.2 Primary, Secondary and Tertiary Measurements

(a) Primary measurements:

- **In primary measurement, required value of a parameter is determined by comparing it directly with known reference or standard.**

Examples:

(i) Measurement of time by counting the number of strokes of a clock.

(ii) Matching of two lengths, when determining the length of an object with a ruler.

(iii) Matching of two colours, when judging the temperature of a red-hot steel.

(b) Secondary measurements:

- Secondary measurements are defined as, ***"the indirect measurements involving one conversion or translation".***

Examples:

(i) Pressure measurement by manometers.

(ii) Temperature measurement by mercury-in-glass thermometers.

(c) Tertiary measurements:

- Tertiary measurements are defined as, ***"the indirect measurements involving two conversions".***

Examples:

(i) Measurement of pressure by a Bourdon tube pressure gauge.

(ii) Measurement of temperature of an object by thermocouple:

  o Refer Fig. 1.2. The primary signal (temperature of object) is transmitted to a thermocouple, which generates a voltage (secondary signal) which is a function of the temperature (first translation/conversion).

  o This generated voltage is then supplied to a voltmeter as input through a pair of wires. The second conversion is then conversion of voltage into length i.e. displacement of pointer over scale, which can be read by observer.

  o In this way, the tertiary signal is transmitted to the brain of observer.

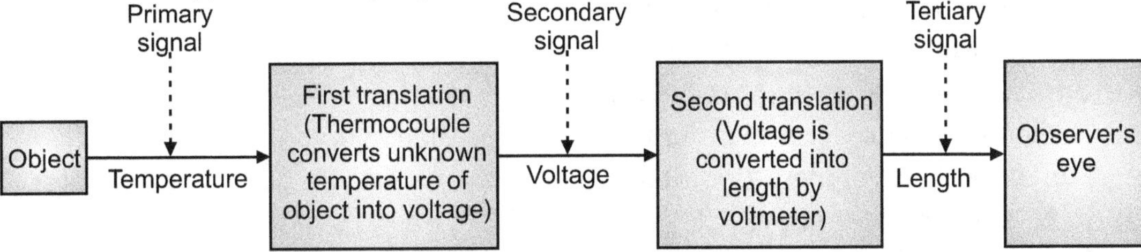

Fig 1.2: Tertiary Measurement: Measurement of Temperature by Thermocouple

1.6 CLASSIFICATION OF STANDARDS

- All the measuring instruments are calibrated at the time of manufacturing against a measurement standard.
- There are four types of standards:

1. International standards:

- These are internationally agreed standards and are maintained at 'International Bureau of Weights and Measures' in Paris.
- They are periodically evaluated and checked by absolute measurement of fundamental units of physics.

2. Primary standards:

- They are maintained at National standard laboratories in different countries.
- They are used to calibrate and verify secondary standards. They are also termed as reference standards, due to their high accuracy.
- These standards are never used outside the National laboratories.

3. Secondary standards:

- They are maintained as reference standards in industries for the purpose of measurement and calibration of working standards.
- They are periodically sent to National standard laboratory for calibration.

4. Working standards:

- They are used to check and calibrate the instruments in laboratory.

1.7 MEASURING INSTRUMENTS

- Measuring instruments are the requirements of accurate and precise measurement.
- The instruments should work efficiently to give quantitative information.
- At the same time, it should take care of comfort and safety limitations of the human operator, while setting up and using it.

1.7.1 Classification of Instruments

- There are number of ways, by which, instrument can be classified.

1. According to method of contact:

(a) Contact type:

- o A contacting type instrument is one, which makes a physical contact with the quantity/ object, whose magnitude is to be determined.
- o **For example:** A thermometer used to know fever of sick person, is inserted in mouth for some time. It senses the temperature by making actual physical contact.

(b) Non-contact type:

- o Non-contact type of instrument is placed remote i.e. away from the measuring medium.
- o **For example:** Optical pyrometer.

2. According to mode of operation:

(a) Manually operated instrument:

- o It requires the service of an operator.
- o **For example:** Null type potentiometer.

(b) Automatically operated instrument:

- o These devices do not require manual help for their running.
- o **For example:** Temperature measurement by mercury in glass thermometer.

3. According to source of energy:

(a) Self-operated instrument or Passive instruments:

- o These instruments do not require any external power in performing its function i.e. the input signal is sufficient to operate the instrument and display the output reading. These are called as **passive instruments**.
- o **For example:** Dial indicator used to measure displacement.

(b) Power operated instrument or Active instruments:

- o These equipments require external power source, such as, electricity, compressed air etc. for their operation. The input signal is not sufficient to operate and display the output reading. These are called as **active instruments**.
- o **For example:** Displacement measurement by LVDT.

4. According to arrangement of physical parts:

(a) Self-contained instrument:

- These instruments contain the whole arrangement of their physical parts including display unit at one place i.e. it is a physical assembly of all parts at one place.
- **For example:** A simple U-tube manometer.

(b) Remote indicating instruments:

- In these instruments, the primary sensing instrument is located at a sufficiently long distance from secondary indicating element.

5. According to output display:

(a) Null type instrument:

- In these types of instruments, deflection is made zero by applying an effect opposing to that generated by the measured quantity.
- **For example:** Pan balance, Dead weight gauge.

(b) Deflection type instrument:

- In this type of instruments, the measurement is done by observing the relative displacement between pointer and dial.
- **For example:** Platform weighing scale, Bourdon tube pressure gauge.

6. According to applications:

(a) Mechanical Instruments

(b) Electrical Instruments

(c) Electronic Instruments

1.7.2 Comparison between Deflection and Null Type Instruments

Sr. No.	Comparative Point	Deflection Type Measuring Instruments	Null-output Type Measuring Instruments
1.	**Working principle**	In this type of instruments, the measurement is done by observing the relative displacement between pointer and dial.	In this type of instruments, deflection is made zero by applying an effect opposing to that generated by the measured quantity.
2.	**Construction**	Simple.	Complex.
3.	**Accuracy**	Less.	More.
4.	**Sensitivity**	Less.	More.
5.	**Response**	Fast.	Slow and Poor.
6.	**Examples**	Platform weighing scale, Bourdon tube pressure gauge etc.	Pan balance, Dead weight etc.

1.8 ERROR

- **Error** is defined as, ***"the difference between measured value and true value of the measured quantity".***
- Errors are unavoidable in any measurement system, i.e. no measurement is free from error.
- The accuracy of any measurement system is measured in terms of errors. It is never possible to measure the true value or exact value of quantity, but only possible to give the best measured value of quantity.
- **Static error** is defined as, *"the difference between best measured value and true value of quantity."*

- The quality of measurement is specified by relative static error.
- **Relative static error** is defined as, *"the ratio of absolute error and measured value"*.

$$\text{Relative static error} = \frac{\text{Absolute error}}{\text{Measured value}}$$

1.9 CLASSIFICATION OF ERRORS

- Errors in measured values may be introduced either due to:
  (i) Characteristics of measuring instruments, or
  (ii) Process of measurement influenced by changing environmental conditions.
- Errors are classified into systematic errors and random errors.

(A) Systematic Errors:

- They include,
  1. Instrumental error
  2. Environmental error
  3. Operational error
  4. Translation and signal transmission error
  5. Observational error.

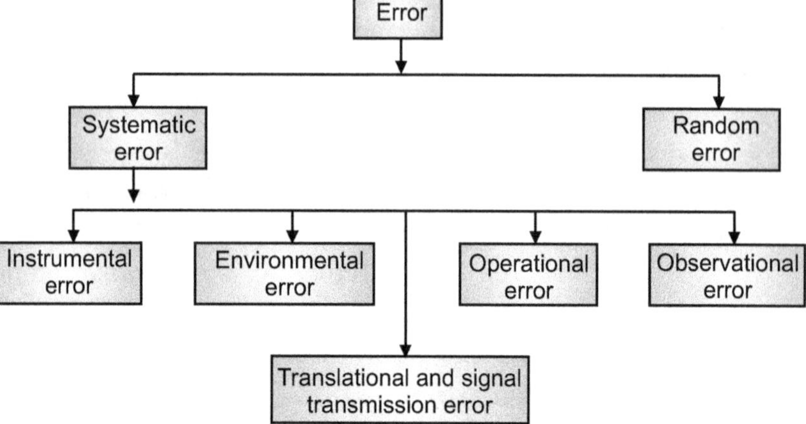

Fig. 1.3: Classification of Errors

1. **Instrumental errors:** These errors are caused due to,
   (i) Improper design and construction of instrument.
   (ii) Improper selection of instrument.
   (iii) Poor maintenance and careless handling of instrument.
   (iv) Mechanical friction, wear, backlash and hysteresis.
- These errors can be reduced by proper design, construction, maintenance and careful handling of instrument.

2. **Environmental errors:** These errors are caused due to,
   (i) Different environmental conditions at the time of manufacturing and at the time of use.
      For example: Temperature, pressure, humidity etc. The effect of temperature is predominant.
   (ii) Presence of strong electromagnetic or electrostatic field, vibrations etc. at the time of use of instrument.
- **These errors can be reduced,**
   (i) By using the instrument under the same condition, at which, it was assembled and calibrated.
   (ii) By providing suitable air conditioning, shielding, proper earthing and spring mounting.
   (iii) By making completely new calibration to suit the local conditions.
   (iv) By measuring the deviations in the local atmospheric condition from the calibration condition and then, by applying suitable correction to the instrument readings.

3. **Operational errors:** These errors are caused due to misalignment or improper method of operation.
   Examples:
   (i) Thermometer will not show proper reading, if its thermal bulb is not installed properly.
   (ii) Flow meter will give wrong reading, if it is installed immediately after a bend or valve (pipe fitting) in the piping.
- **These errors can be reduced by** adopting the best suitable method of measurement and by taking more number of measurements for same reading.

4. Translational and Signal transmission errors: They may arise in the conditions such as,

   (i)   The instrument may not sense or translate the measurement effect with complete accuracy.
   (ii) Non-capability of instruments.
   (iii) The transmitted signal becomes faulty due to distortion by resonance, loss, leakage or on being consumed or absorbed within the communication signal.
   (iv) Due to unwanted disturbances, such as, noise or small waves (ripples) etc.

- **These errors can be reduced by** calibration and by monitoring at one or more points along the transmission path.

5. Observational errors: These errors are caused due to,

   (i)   Parallax, i.e. line of vision is not normal to the plane of scale.
   (ii) Inaccurate conversion of unit.
   (iii) Inaccurate estimate of average reading.
   (iv) Wrong scale reading and wrong recording of data.

- **These errors can be minimized by:**

   (i)   Taking great care, while reading and recording the data.
   (ii) Two, three or more number of readings should be taken for quantity under measurement.

(B) Random Errors (Residual Errors):

- Even if, all the systematic errors are accounted and controlled, it is found that, experimental results show variation amongst the number of readings taken.
- This is due to certain factors influencing the measurement, which are not known to us.
- These errors are accidental in nature. Therefore, they cannot be traced and minimized.
- These errors are called as **random errors** or **residual errors** or **system interaction errors.**
- **For example:**

   (i)   Reading shown by hand tachometer would vary with pressure, with which, it is pressed against the shaft.
   (ii) An obstruction type flow meter (orifice meter) may partially block or disturb the flow; therefore, the flow rate shown by the meter may not be accurate.

- These errors are very small and independent. They are mainly due to non-assignable causes or factors, such as, spring hysteresis, friction, noise, improper setting etc.

1.9.1 Difference between Systematic and Random Errors

Systematic Errors/Avoidable Errors	Random Errors/Unavoidable Errors
1. These errors are consistent and repetitive in nature and are of similar form.	1. These errors are non-consistent and accidental in nature.
2. They occur due to improper conditions or methods of measurements.	2. They are inherent in the measuring system or measuring instruments. It is assumed that, occurrence of random error is bound to be there.
3. Except human error, all other systematic errors can be controlled in magnitude and sense.	3. Specific causes, magnitude and sense of random errors can not be determined from the knowledge of measuring system or conditions at the time of measurement.
4. These errors can be identified and reduced, if attempts are made to analyze them.	4. These errors can not be eliminated, but the results obtained can be corrected.

contd. ...

5. They include errors caused due to (i) Variation in environmental conditions, (ii) Improper calibration, (iii) Incorrect contact pressure or stylus pressure applied, (iv) Parallax, (v) Improper location of measuring instrument.	5. They include errors caused due to (i) Small variations in the positions of standard and the workpiece, (ii) Slight displacement of lever joints in the measuring instruments, (iii) Transient fluctuations in friction in measuring instrument, (iv) Operator error due to incorrect reading of engraved scale on readout display device.
6. Statistical methods do not operate on the systematic errors.	6. Statistical methods only operate on the random errors.

1.10 CHARACTERISTICS OF INSTRUMENTS

- Measurement is basically used to monitor a process or operation and further, it can be used for controlling process.
- Thus, for measurement, it is important that, **"how well a measuring instrument is performing its various functions"**.
- This can be determined by **"knowing how closely the instrument output is reflecting with the change in variable that is being measured"**.
- The characteristics of instruments can be divided into two categories:
  1. **Static characteristics.**
  2. **Dynamic characteristics.**

1.10.1 Static Characteristics

- When the instrument is used to measure an input, which is not varying with time, then the behaviour of instrument is called as **static behaviour**.
- Static characteristics can be defined as, *"a set up criteria, which provides meaningful description of quality measurement under static conditions."*
- **Static characteristics** are:

Desirable static characteristics	Undesirable static characteristics
(i) Accuracy	(i) Drift
(ii) Precision	(ii) Static error
(iii) Sensitivity	(iii) Dead zone
(iv) Reproducibility	(iv) Non-linearity
(v) Linearity	(v) Hysteresis
(vi) Readability	(vi) Threshold

1.10.2 Dynamic Characteristics

- When an instrument is used to measure an input varying with time, then the behaviour of instrument is called as **dynamic or transient behaviour**.
- In case of industrial, aerospace and biological applications of measuring system, the inputs are not static, but are dynamic in nature i.e. the inputs vary with time.
- The input varies from instant to instant and therefore, the output also varies.
- This behaviour of system under such conditions is described by a term, **"dynamic response"**.
- **Dynamic characteristics** are:

Desirable dynamic characteristics	Undesirable dynamic characteristics
(i) Speed of response	(i) Measuring lag
(ii) Fidelity	(ii) Dynamic error
	(iii) Overshoot

1.11 ACCURACY

- Accuracy of measurement means conformity to true value.
- **Accuracy** is defined as, ***"the agreement of result of measured quantity with its true value."***
- In other words, it is closeness of an instrument reading to approach the true value of quantity being measured.
- It is influenced by static error, dynamic error, drift, reproducibility and non-linearity.
- All devices and instruments are classified and designated into different grades or classes depending upon the accuracy of product.
- Mathematically, accuracy on the basis of true value (V_t) is given by,

$$\text{Accuracy} = \frac{V_{max} - V_t}{V_t} \times 100\ \%$$

or

$$\text{Accuracy} = \frac{V_{min} - V_t}{V_t} \times 100\ \%$$

where, V_{max} = Maximum value of reading taken by instrument

    V_{min} = Minimum value of reading taken by instrument

- This is the representation of accuracy on the basis of true value.

1.12 PRECISION

- **Precision** is defined as, ***"the ability of the instrument to reproduce consistent reading for a constant input."***
- Therefore, we say that, Precision defines repeatability of the process.
- It has no meaning for single measurement. However, it refers to the degree of agreement, within a group of measurements.
- High precision means a tight cluster of repeated results, while low precision means a wide scattering of results. But, high precision does not indicate high accuracy all the time.
- Difference between accuracy and precision can be illustrated with the help of following Fig. 1.4.
- The arrangement shown corresponds to the **shooting range**, where one is asked to strike a target represented by a center circle.
- These centre circles represent the true value, whereas the various marks 'X' indicate the results achieved by the striker.

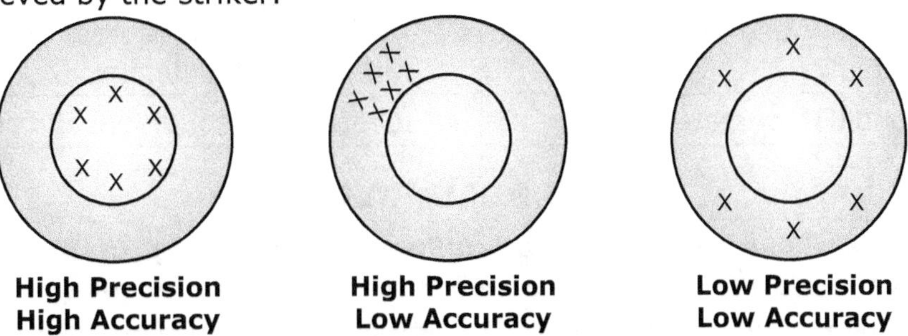

Fig. 1.4: Examples Illustrating Difference between Accuracy and Precision

1.12.1 Difference between Accuracy and Precision

Accuracy	Precision
1. Accuracy is defined as," *the agreement of result of measured quantity with its true value."*	1. Precision is defined as, *"the ability of the instrument to reproduce consistent reading for a constant input."*
2. High accuracy cannot be obtained with low precision.	2. High precision can be obtained with low accuracy.

contd. ...

3. High accuracy guarantees high precision.	3. High precision cannot guarantee high accuracy.
4. Accuracy is concerned with true value.	4. Precision has no concerned with true value.
5. It is difficult and expensive to achieve good accuracy.	5. It is easy and cheaper to achieve good precision than accuracy.
6. Accuracy is determined by proper calibration.	6. Precision is determined by statistical analysis.

1.13 REPEATABILITY

- **Repeatability** is defined as*, "the ability of the measuring instrument to repeat the output readings consistently, when the same quantity is measured number of times under identical conditions (i.e. same observer, same instrument and same environmental conditions within short intervals of time)."*

- A measurement is said to be repeatable, when variation in the measurement is smaller than the agreed limit.

- Repeatability is also known as inherent precision of a measuring instrument.

1.14 REPRODUCIBILITY

- Reproducibility is defined as*, "the closeness of agreement between the results of measurement of the same quantity, when individual measurements are carried out by different observers, by different methods, using different instruments under different environmental conditions."*

- Reproducibility shows variability of measurement system caused due to difference in operator behaviour.

1.15 DIFFERENCE BETWEEN REPEATABILITY AND REPRODUCIBILITY

Repeatability	Reproducibility
1. Repeatability is the variation in measurement taken by single person or instrument on the same item and under the same condition.	1. The closeness of agreement between independent results obtained by different observers, using different measuring instruments, while measuring same quantity, but under different conditions.
2. The conditions of use of instrument remain unchanged.	2. The conditions of test are allowed to vary within specified limits.
3. It refers the condition of stability.	3. It refers the conditions of consistency.

1.16 RELIABILITY

- **Reliability** is defined as, *"the probability of measuring instrument, that it will perform its function for a given specific time period under given conditions."*

- It is affected by manufacturing method, quality of maintenance and skill of operator.

1.17 SENSITIVITY

- **Sensitivity** is defined as, *"the ability of measuring instrument to respond to very small variation in the input (quantity being measured)."*

- **Static sensitivity** of an instrument or an instrumentation system is defined as, *"the ratio of magnitude of change in output signal to the magnitude of change in input signal."*

Mathematically,

$$\text{Static sensitivity, } K = \frac{\text{Change in output signal (response)}}{\text{Change in input signal (being measured)}}$$

- If δq_o is the very small change in output and δq_i is the very small change in input, then, K is expressed as,

$$K = \frac{\delta q_o}{\delta q_i}$$

- The following two graphs (a) and (b) shows the variation of sensitivity with the input.

- In graph (a), sensitivity $\left(K = \dfrac{\delta q_o}{\delta q_i}\right)$ is constant over the entire range of instrument. Therefore, the calibration curve appears to be **linear**. In this case, the sensitivity of instrument can be defined as **slope of calibration curve.**

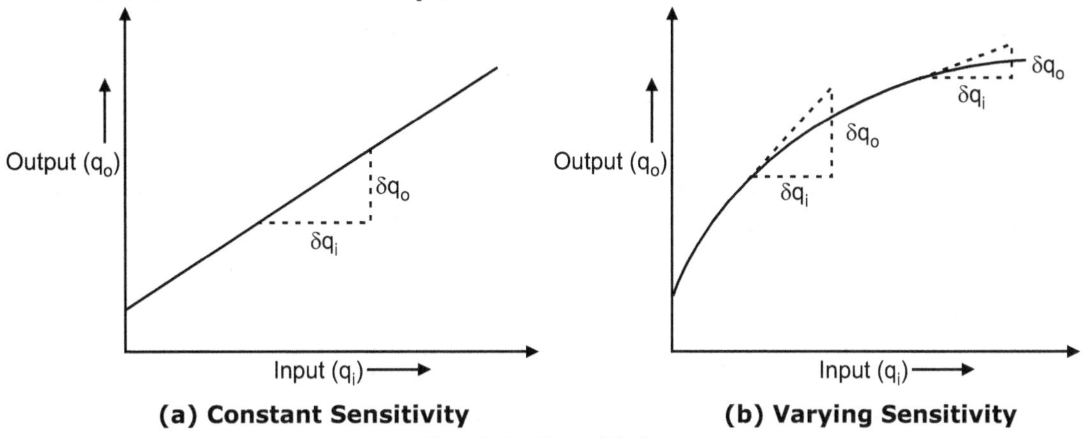

(a) Constant Sensitivity   **(b) Varying Sensitivity**

Fig. 1.5: Sensitivity

- In graph (b), the sensitivity varies with the input. Therefore, the graph appears to be curve instead of a normal straight line. Therefore, the sensitivity is specified as **non-linear calibration curve**.

- Too sensitive instrument may lead to drifts due to thermal or other effects and output indications will be less repetitive (precise).

- Sensitivity has wide range of unit and it depends upon the measurand and the measuring instrument.

- **For example:** The operation of a **Resistance Thermometer** depends upon change in resistance (output) due to change in temperature (input). Therefore, its **sensitivity** is expressed in terms of **Ohms per degree Celsius**.

- Consider a system comprising of different elements connected in series or cascades, then overall sensitivity is calculated in the following manner.

$$K_1 = \frac{q_1}{q_i} , \quad K_2 = \frac{q_2}{q_1} , \quad K_3 = \frac{q_o}{q_2}$$

Fig. 1.6: Overall Sensitivity

$\therefore$ **Overall sensitivity,**

$$K = \frac{q_o}{q_i}$$

$$K = \frac{q_1}{q_i} \times \frac{q_2}{q_1} \times \frac{q_o}{q_2}$$

$$\mathbf{K = K_1 \times K_2 \times K_3}$$

1.18 LINEARITY AND NON-LINEARITY

Linearity:

- **Linearity** is defined as, **"the ability of the instrument to give output, which is linearly proportional to input."** Refer Fig. 1.7 (a).

- This is a desirable property of an instrument.

Importance of Linearity:

- When a scale reading is to be converted to a value corresponding to the measured input, linearity is important, because one has to multiply scale reading by a fixed constant to get measured input.
- If the non-linearity appears, then the reading has to be obtained by computing non-linear calibration equation.
- Closeness of calibration curve to the specified straight line is called as **linearity of instrument.** Linearity is also called as **degree of straight-line relationship**.
- **For example:** The resistance used in a potentiometer should vary linearly with displacement of the sliding contact, so that, the obtained output voltage is directly proportional to the displacement. Any departure from linearity would result into error in the read-out system.
- If the instrument calibration curve is not a straight line, it should not be concluded that, such an instrument with non-linear behaviour is inaccurate. However, it is desirable to have the measuring system of a linear characteristic.
- Linearity is measured in terms of percent linearity. It is given by,

$$\text{Percent linearity} = \frac{\text{Maximum deviation}}{\text{Full scale reading}} \times 100$$

- Linearity can be expressed by the straight-line equation.

$$y = m \cdot x + C$$

where,

$$y = \text{output} = q_o$$
$$x = \text{input} = q_i$$
$$m = \text{slope of line}$$
$$C = \text{'Y' intercept (if any)}$$

Therefore,

$$q_o = m \times q_i + C$$

If $C = 0$, then

$$q_o = m \times q_i$$

∴

$$m = \text{slope} = \frac{q_o}{q_i}$$

Non-Linearity:

- Non-linearity is referred as any departure from straight line relationship.
- **Non-linearity of instrument** is defined as, ***"deviation of calibration curve from the specified straight line."*** Refer Fig. 1.7 (b).
- It may be due to non-linear elements in the measurement device, mechanical hysteresis, creep etc.

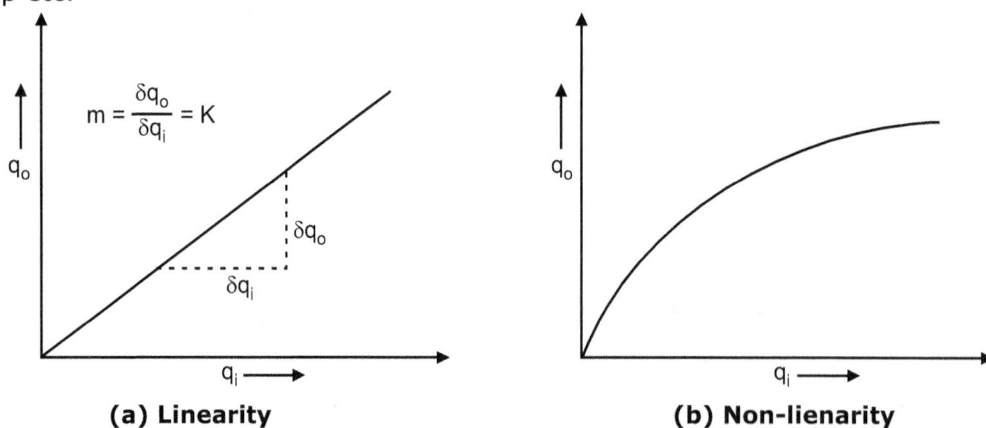

(a) Linearity   (b) Non-lienarity

Fig. 1.7: Linearity and Non-Linearity

1.19 DRIFT

- The output of a measuring system should change with change in input, but, sometimes there are gradual variations in the output over period of time, which are not related to changes in input. This situation is called as **drift**.
- **Drift** is defined as, ***"the undesired change or a gradual variation in the output over period of time, which is not related to changes in input, operating condition or load"***.
- Drift occurs very slowly and it can be checked by periodic inspection and maintenance of instrument.
- **For example:**
  - (a) In case of flow meter, such as, orifice meter, drift occurs due to wear and erosion of orifice plate.
  - (b) In case of thermocouple and resistance thermometer, drift occurs due to contamination of metals and change in their atomic structure.

1.19.1 Types of Drift

- Drift can be defined in the following ways:
  1. **Zero drift:** When the instrument calibration gradually drifts due to slippage, permanent set or due to undue warming up of electronic tube circuit etc., the drift is called as **zero drift.** Refer Fig. 1.8 (a).
  2. **Span drift:** If there is proportional change in output indication, all along the straightway, the drift is called as **span drift.** Refer Fig. 1.8 (b).
  3. **Zonal drift:** If the drift occurs only in a particular zone or span or proportion of a span of an instrument, the drift is called as **zonal drift.**

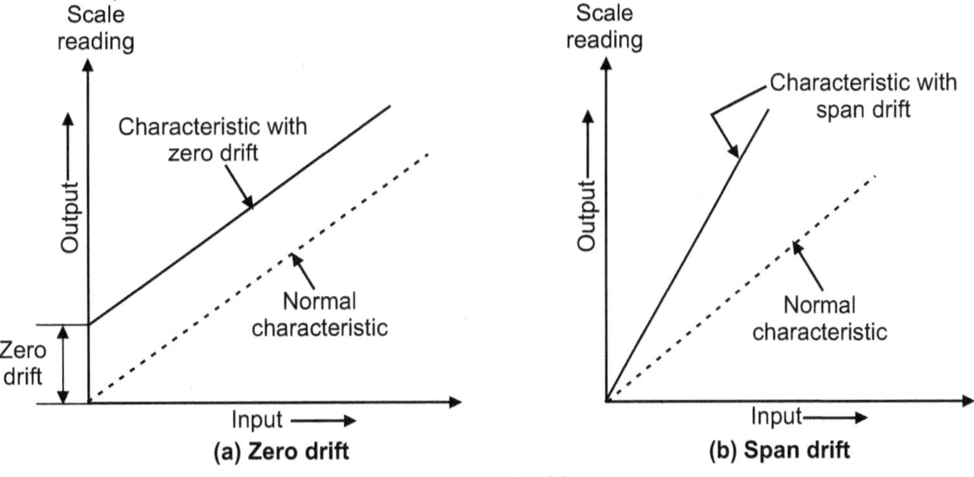

Fig. 1.8: Drift

1.19.2 Causes of Drift

- Drift may occur due to,
  1. Wear and tear.
  2. Development of high stresses at some parts of measuring instrument.
  3. Contamination of primary sensing element.
  4. Change in atomic structure of metals used in measuring instrument.
  5. Effect of electrostatic and electromagnetic field.
  6. Effect of vibration.

1.19.3 Prevention of Drift

1. Drift can be prevented by periodic inspection and maintenance.
2. The effect of electrostatic and electromagnetic field can be prevented from affecting measurement with the use of proper shielding.

3. Effect of mechanical vibration is avoided by having proper mounting.

4. The change in temperature during the measurement process should be preferably avoided.

1.20 HYSTERESIS

- If a device is used to measure any parameter, first for increasing values of measurand (loading) and then for decreasing values (unloading) and we get two output readings for same input, then this situation is called as **Hysteresis.**

- **Hysteresis** is defined as, *"the non-coincidence of loading and unloading curves."*

- Hysteresis is determined by **the maximum difference in any part of output readings.**

- This effect of hysteresis may occur in any physical, chemical or electrical phenomenon.

- **Consider an instrument,** which has **no friction due to sliding parts.**

- If input of this instrument is slowly increased from zero value up to full scale and then slowly decreased from full scale to zero value, its output will vary as shown in Fig. 1.9 (a). This non-coincidence of output (when input is increased and decreased) is due to **internal friction or hysteretic damping**. Thus, we obtain two different values of output for same input under increasing or decreasing condition.

- In case of the instruments, which are used on both sides of zero i.e. input is applied on both, positive and negative side; the variation of output will be as shown in Fig. 1.9 (b).

- In case of instruments, which do not have internal friction, but have **external sliding friction i.e. constant coulomb friction**. The input-output relationships are as shown in Fig. 1.9 (c) and (d).

- Combining all the effects for an instrument, the input-output relationship will appear as shown in Fig. 1.9 (e).

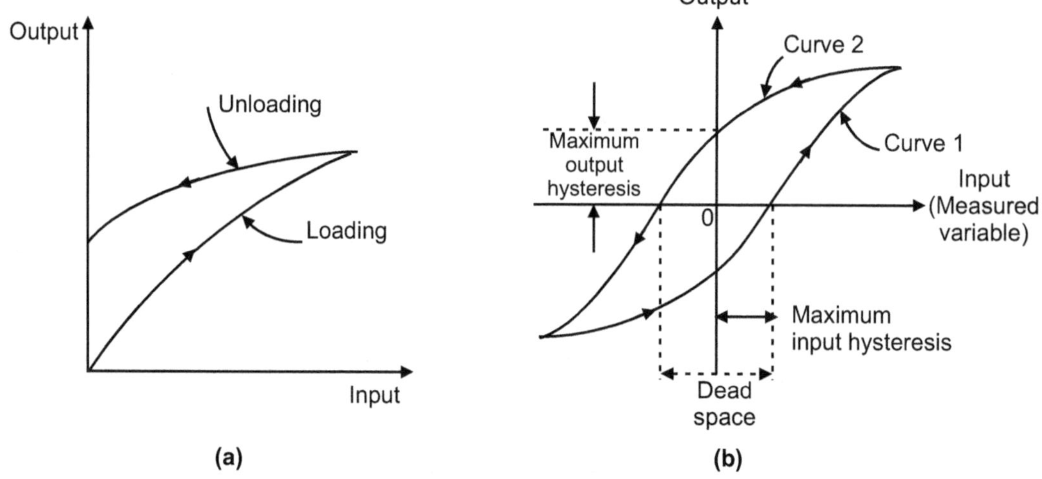

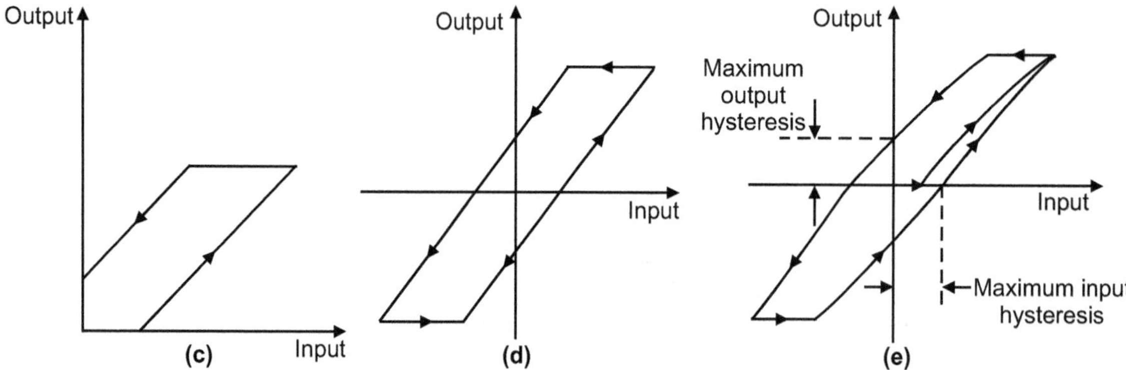

Fig. 1.9: Hysteresis

1.20.1 Causes of Hysteresis

- Hysteresis may cause due to presence of irreversible phenomenon such as,
  1. Mechanical friction.
  2. Slack motion in bearings.
  3. Elastic deformation.
  4. Thermal effects.
  5. Magnetic effect.
  6. Due to heating and cooling effect in electronic system under the condition of rising and falling input.

1.21 BACKLASH

- **Backlash** is defined as, *"the maximum distance or angle, through which, any part of mechanical system may be moved in one direction without applying force or motion to the next part in a mechanical system".*

1.22 DEAD ZONE

- **Dead zone** is defined as, *"the largest range, through which, an input signal can be varied without any change in output of instruments."*
- It is an undesirable property.
- **For example:** The input applied to the instrument may not be sufficient to overcome the friction and therefore, the output will not move at all. It will move only when, input is such that it produces a driving force, which can overcome the friction.

1.22.1 Factors Causing Dead Zone

1. Friction involved in the components of instrument.
2. Play between the mating components of instrument.
3. Backlash present in the instrument.
4. Hysteresis in the instrument

1.23 DEAD TIME

- Dead time is defined as, *"the time required for measurement system to begin to respond to change in the measurand."*

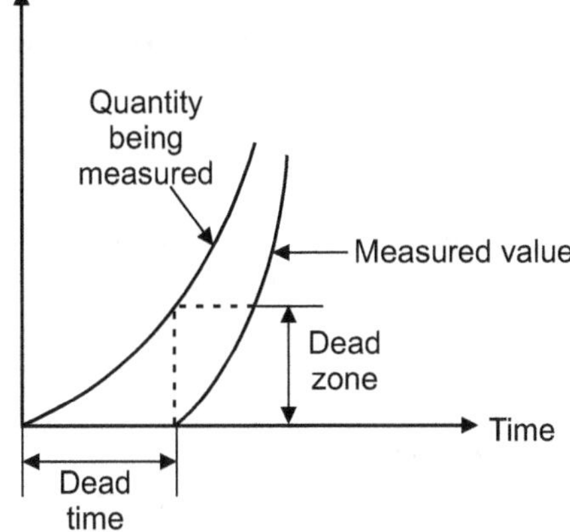

Fig. 1.10: Representation of Dead Zone and Dead Time

1.24 THRESHOLD

- Threshold of instrument is defined as, *"the minimum value, below which, no change in output value can be detected, when input of an instrument is increased gradually from zero."*

- When the input signal to an instrument is gradually increased from zero, there will be some minimum value of input, before which, the instrument will not detect any output change. This minimum value of input is called as threshold of instrument.
- Thus, **threshold** can be defined as, *"the minimum value of input, which is necessary to cause a detectable change from zero output."*
- In a digital system, it is the input signal necessary to cause one least significant digit of the output reading to change.

1.25 RESOLUTION

- Resolution is defined as, ***"the small increment in the input quantity from a non-zero value being measured, which can be detected by an instrument."***
- When input signal is increased from non-zero value, it is observed that, the output does not change, until a certain amount of input increment is exceeded. This increment is called as **resolution or discrimination**.
- Thus, resolution is also defined as, ***"the smallest measurable change in input, for which, there will be a change in output."***

1.25.1 Difference between Threshold and Resolution

Threshold	Resolution
1. Threshold is defined as, *"the minimum value, below which, no change in output value can be detected, when input of an instrument is increased gradually from zero."*	1. Resolution is defined as, *"the small increment in the input quantity from a non-zero value being measured, which can be detected by an instrument."*
2. It starts from zero value.	2. It starts from non-zero value.
3. Threshold defines the minimum value of input, which is necessary to cause a detectable change in value of zero output.	3. Resolution defines the smallest measurable change in input, for which, there will be a change in output.

1.26 DYNAMIC CHARACTERISTICS

1. Speed of response:
- Speed of response is defined as, ***"the value of rapidity (rapidness), with which, an instrument responds to the change in the quantity being measured".***

2. Measuring lag:
- Measuring lag is defined as, ***"the retardation or delay in the response of an instrument to measure a given quantity."***
- This is caused by conditions such as capacitance, inertia, resistance etc.

3. Fidelity:
- Fidelity is defined as, ***"the degree, to which, a measurement system indicates change in the measured quantity without any dynamic error."***
- Fidelity is also defined as, *"the degree of closeness, with which, the system indicates or records the signal, which is impressed upon it."*
- Fidelity refers to the ability of system to reproduce the output in the same form as the input.
- **For example:** If input signal is sine wave, then with 100% fidelity, the output signal will also be a sine wave. Refer Fig. 1.11.

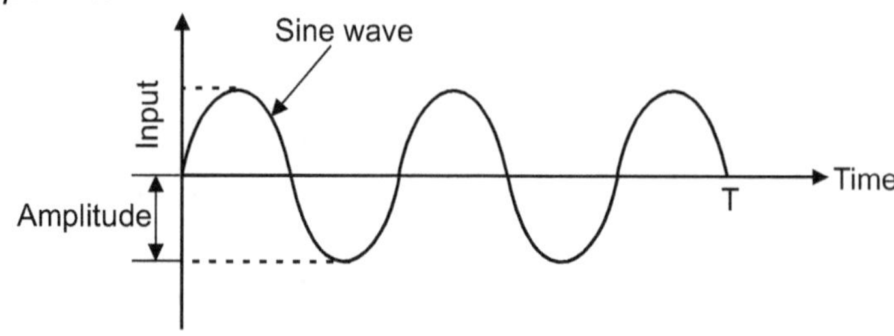

Fig. 1.11: Fidelity

4. Dynamic error:

- Dynamic error is defined as, ***"the difference between the true value of quantity (under measurement), changing with time and the value indicated by measurement system."***
- Here, the static error is assumed to be zero. It is also called as ***measurement error.***

5. Overshoot:

- Due to mass and inertia, the pointer of the instrument (moving member) does not immediately come to rest in the final deflected position i.e. the pointer goes beyond the steady state. This is a situation of overshoot. Refer Fig. 1.12.

- **Overshoot** is defined as, ***"the maximum amount, at which, the pointer moves beyond the steady state."***

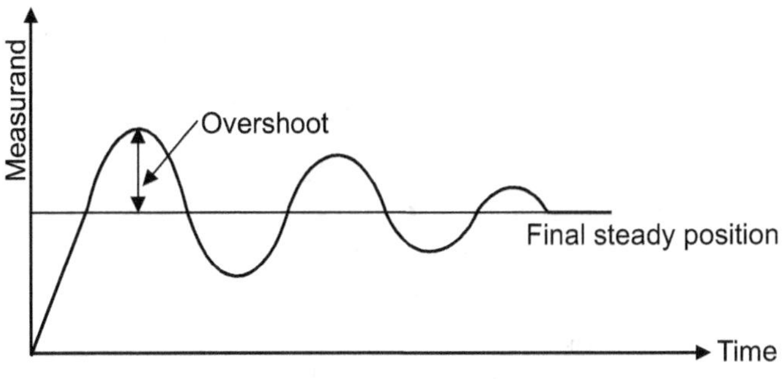

Fig. 1.12: Overshoot

1.27 CALIBRATION

- Calibration is the process of establishing a reliable measuring instrument.
- Calibration is usually carried out by making adjustment, such that, read out device produces zero output for zero input and similarly, it should display output equivalent to known input.
- Calibration is defined as, ***"the process of framing of scale of instrument by applying some standard."***
- Calibration is said to be a **pre-measurement** process, carried out at the time of manufacturing the measuring instrument.
- It is the entire procedure laid down (specified), which includes marking, adjusting, checking etc., so that, the reading obtained will conform to the agreed and accepted standard.
- This involves the comparison of instrument to be calibrated with primary standard, secondary standard or a known input standard.

    For example: Flow meter can be calibrated by any one of the following methods.

    (i) Comparing it with a standard flow measurement facility available at the National Bureau of Standards (primary standard).

    (ii) Comparing it with another flow meter (a secondary standard), which has been already compared with a primary standard.

    (iii) By direct comparison with primary measurement system involving the activities, such as weighing certain amount of water in the tank and recording the time required for this quantity to flow through the meter.

- Calibrated devices permit a manufacturer to produce a good quality product (such as flowmeter) with desired specification.
- Calibration requires highly skilled operator, standard environmental conditions and availability of good standard, which can be used as reference.

1.27.1 Necessity of Calibration

- **Calibration of all instruments** is very important, as it offers the opportunity to check the instrument against the known standard.
- It is tried to perform 'calibration' under those environmental conditions, which are very close to the conditions, under which, actual measurements will be carried out.
- Even if, the sensing system and measuring system are different, then also, it is necessary to calibrate the system, so that, the error producing components are eliminated or rectified.
- Calibration is done by applying standards in such a way that, 'Output read-out device shows zero output for zero input' and similarly, 'It should show an output equivalent to the known input'.

1.27.2 Recalibration

- **Recalibration** is carried out, if the device has been repaired or modified.
- Consistent use of measuring instruments affects their accuracy. If we need to maintain the accuracy, the measuring instruments must be checked and recalibrated.
- The periodic cycle or schedules of such calibration process depend upon the severity of use, surrounding environmental conditions, required accuracy etc.

1.27.3 Factors Governing Calibration

- Calibration depends upon following factors:
  1. Desired accuracy.
  2. Range of instrument.
  3. Desired sensitivity.
  4. Applications of instruments.
  5. Environmental conditions.

1.27.4 Factors to Be Considered at The Time of Instrument Calibration

1. Calibration should be carried out with the instrument in the same position (horizontal, vertical etc.) and in same environmental conditions, under which, it is required to operate.

2. Instrument should be calibrated with the values of measurand impressed in both, increasing and decreasing order.

3. Calibration must be performed periodically to taste the validity of performance of the measuring instrument or device or the measuring system.

1.28 RANGE OF THE INSTRUMENT

- **Range** is defined as, *"the region between the limits, within which, an instrument is desired to operate for measuring, indicating or recording a physical quantity."*
- **Example:** If a thermometer is used to measure temperature between 0°C to 100°C, then the range will be expressed as **"Range = 0°C to 100°C"**.

1.29 SPAN OF THE INSTRUMENT

- **Span** is defined as, *"the algebraic difference between upper and lower measurement limits of any instrument."*
- **Examples:**
  (i) If a thermometer is used to measure temperature between 0°C to 100°C, then the span will be expressed as,

$$\textbf{Span} = \text{Highest calibration value} - \text{Lowest calibration value}$$
$$= 100°C - 0°C$$
$$= \textbf{100°C}$$

  (ii) If a voltmeter is used to measure voltage between 25 volts to 75 volts, then the span will be expressed as,

$$\textbf{Span} = \text{Highest calibration value} - \text{Lowest calibration value}$$
$$= 75\,V - 25\,V$$
$$= \textbf{50 Volts}$$

Note: If X is the lowest point of calibration and Y is the highest point of calibration on the scale of the instrument, then,

(a) Range will be from 'X' to 'Y'.

(b) Span will be equal to (Y – X).

1.30 FACTORS TO BE CONSIDERED WHILE SELECTING A SUITABLE MEASURING INSTRUMENT

1. Quantity to be measured, e.g. Length, Displacement, Pressure, Temperature etc.
2. Range of measuring instrument.
3. Precise measurements.
4. More accuracy.
5. Sensitivity and Readability of instrument.
6. Computability of instrument.
7. Easy to operate.
8. Low capital cost.
9. Easy for maintenance at low cost.
10. Simple in design with elaborated constructional details.

1.31 ESSENTIAL CHARACTERISTICS FOR AN ACCURATE MEASURING INSTRUMENT

1. Measuring instrument should possess the desired requirements and consistent 'accuracy'.
2. Every important source of possible error should be known and the instrument itself should have adjustments to eliminate the errors.
3. If error cannot be eliminated completely, then at least, the measuring instrument should be capable of reducing it to a very small value.
4. Less response to avoid influence of varying atmospheric conditions.

1.32 TRANSDUCER

- **Transducer** is a device, which senses the measurand (quantity to be measured) as input signal and converts it into an output signal to indicate measured value.

For example:

- Input to the transducer could be pressure, acceleration or temperature and output of transducer may be displacement, voltage or change in resistance. It depends upon the transduction element and working principle used.
- Therefore, transducer can be assumed to be made of:
  1. **Primary sensing element (Detector):**
     - It is the part of transducer, which **responds to a physical phenomenon or change in the physical phenomenon.**
  2. **Transduction element:**
     - It **transforms the output of primary sensing element** into **some another signal**. It is also referred as *secondary transducer.*

1.33 CLASSIFICATION OF TRANSDUCER

- Transducers may be classified in various ways as enlisted below.
  1. Mechanical Transducer and Electrical transducer.
  2. Primary and Secondary Transducers.
  3. Active and Passive Transducers.
  4. Analog and Digital Transducers.
  5. Transducers and Inverse Transducers (Input and output transducers).

1.34 MECHANICAL TRANSDUCERS AND ELECTRICAL TRANSDUCER

1.34.1 Mechanical Transducer

- Mechanical transducer converts **a physical quantity** or mechanical parameter into **an equivalent output of some other form of physical quantity,** which is the measure of quantity under measurement.

- The commonly used mechanical transducers along with their operation (conversion of one form into another) are listed below:

Mechanical Transducer	Operations
1. Contacting spindle: Dial indicator	Displacement to displacement
2. Elastic members: (a) Bourdon tube pressure gauge (b) Bellow and Diaphragm (c) Springs	 Fluid pressure to displacement Fluid pressure to displacement Force to displacement
3. Concentrated mass: (a) Pendulum (b) Manometric liquid	 Force to displacement Fluid pressure to displacement
4. Thermal: Liquid in glass thermometer and Bimetallic thermometers	Temperature to displacement
5. Hydro-pneumatic: (a) Float (b) Orifice plate, Venturimeter, Flow nozzle, Pitot tube.	 Liquid level to displacement Velocity to pressure head

1.34.1.1 Advantages of Mechanical Transducer

1. High accuracy.
2. Low cost.
3. Rigidity and construction.
4. Ability to operate without any external power supply.

1.34.1.2 Limitations of Mechanical Transducer

- Mechanical transducer is **not recommendable for many modern scientific experiments**, **process control, instrumentation** etc. due to following reasons.
  1. Poor frequency response.
  2. Requires large force to overcome mechanical friction, so as, to move pointer on calibrated scale for indication.
- All these drawbacks have been overcomed with the **use of electrical transducer**, but still the **mechanical devices are used as primary detectors or sensors** to collect and extract information from mechanical systems.
- The commonly used **sensing** elements are **spring, diaphragms, bimetallic strip** etc.

1.34.2 Electrical Transducer

- Electrical transducer converts a **physical quantity** or mechanical parameter into **an equivalent electric signal**.
- The electrical signal (output) may be current, voltage or frequency.
- The form of output signal is based upon the method of transduction, whether resistive, capacitive or inductive.

1.34.2.1 Essential Characteristics of an Electrical Transducer

1. Linearity, i.e. linear relationship between the measurand (input) and the measured value (output).
2. Good sensitivity.
3. Wide operating range.
4. Repeatability.
5. Compact with minimum weight and volume, so that, it will occupy less space.

1.34.2.2 Advantages of Electrical Signal

1. Amplification of electrical signal can be done easily.
2. Mass-inertia effects are minimized.
3. Effects of friction are minimized.
4. The output can be indicated and recorded from a remote distance away from the sensing element.
5. The output can be modified to meet the requirements of the indicating or controlling unit.
6. The electrical output can be easily used, transmitted and processed for measurement.
7. The power required to control any electrical or electronic system is very small.
8. The signal can be mixed to obtain any combinations with outputs of similar transducer or control signal.

1.34.3 Comparison between Electrical and Mechanical Transducers

Sr. No.	Electrical Transducers	Mechanical Transducers
1.	These transducers convert the physical quantity into electrical output.	These transducers convert the physical quantity into another physical quantity.
2.	Output is current or voltage proportional to input.	Output is another physical quantity.
3.	Output is easily amplified.	Output cannot be easily amplified.
4.	Examples: Thermistor, Thermocouple etc.	Examples: Bourdon tube pressure gauge, Bellows etc.

1.35 PRIMARY AND SECONDARY TRANSDUCERS

1. Primary Transducer:

- There is only one stage of transduction.
- Primary transducer makes the first contact with the measurand, senses it as a physical parameter (like temperature, pressure) and converts it into suitable and readable physical parameter (like displacement).
- **Examples:**
  (i) Bourdon tube senses pressure and converts it into displacement at free end.
  (ii) Weighing spring converts the weight to be measured in the form of displacement.

2. Secondary Transducer:

- There are two stages of transduction. Here, the primary sensor of the instrument makes the first contact with the quantity to be measured and then converts it into some other suitable form of signal, which represents the measured value.
- **For example:** Measurement of pressure by combined system of Bourdon tube and LVDT.
- Here, the bourdon tube acting as a primary transducer senses the pressure (physical parameter) and converts it into displacement (physical parameters) at its free end.
- The displacement of free end moves the core of a LVDT in vertical direction, which produces an output voltage. This output voltage is proportional to movement of core, hence proportional to displacement of free end. This displacement of free end is the measure of pressure (measurand).

- Thus, in first stage, pressure is converted into displacement in bourdon tube (primary transducer) and in second stage, this displacement is converted into analogous voltage by LVDT (secondary transducer).

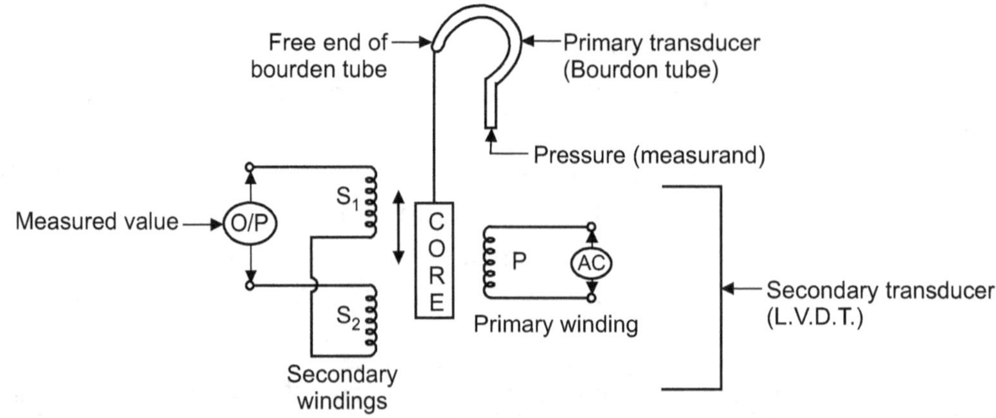

Fig. 1.13: Secondary Transducer

1.35.1 Comparison between Primary and Secondary Transducers

Sr. No.	Primary transducers	Secondary transducers
1.	Most of Primary transducers are sensors.	These are transduction elements.
2.	Directly makes contact with measurand, i.e. quantity being measured.	Connected to the primary transducer.
3.	Since output of primary transducer is given as input to secondary transducer, therefore, working of secondary transducer depends upon primary transducer.	Since output of secondary transducer is not given as input to primary transducer, therefore, working of primary transducer does not depend upon secondary temperature.
4.	Examples: Bourdon tube pressure gauge, Weighing spring etc.	Example: LVDT, RTD, Thermistor etc.

1.36 ACTIVE AND PASSIVE TRANSDUCERS

1. Active Transducers:

- They do not require external power to operate.
- The energy required to generate output signal is obtained from the physical quantity being measured. Therefore, they are known as **self-generating transducers**.
- They generate an equivalent electrical output signal (voltage or current), which is the measure of physical quantity being measured.
- **For example:** Thermocouples for Measurement of Temperature, Photovoltaic Cell for Measurement of Light Intensity, Piezoelectric Crystals for Measurement of Force, Mcleod gauge for Measurement of Pressure etc.

2. Passive Transducers:

- They require an external source of power for their operation. Therefore, they are known as **power operated transducers**.
- The output signal is in electrical form. This output is a measure of variation in resistance, inductance and capacitance.
- **For example:** LVDT, Hot wire anemometer, Pirani Gauge etc.

1.36.1 Comparison between Active and Passive Transducer

Sr. No.	Active transducer	Passive transducer
1.	It operates under energy conversion techniques.	It operates under energy controlling principles.
2.	It does not require an external power.	It requires an external power for its operation.
3.	It is self-generating.	It is externally power operated.
4.	It generates equivalent output signal without any energizing source.	Here, measurand is converted into passive parameter, such as, resistance or capacitance which needs external energy source so as to generate an equivalent output signal.
5.	Examples: Thermocouple, Photovoltaic cell, Mcleod gauge.	Examples: LVDT, Hot wire anemometer.

1.37 ANALOG AND DIGITAL TRANSDUCERS

- The transducer can be classified on the basis of output, i.e. output may be a continuous function of time or the output may be in discrete steps.
  1. **Analog Transducer:**
     - Analog transducer converts the input quantity into an analog output, which is a continuous function of time.
     - **For example:** LVDT, Thermocouple, Thermistor, Strain gauge, Potentiometer etc.
  2. **Digital Transducer:**
     - Digital transducer converts the input quantity into an electrical output, which is in the form of pulses or digitally coded form.
     - **For example:** Digital LVDT (here, a scale is constructed to show the linear position of a movable object with 5 digits indication), Rotary (Shaft) encoder.

1.37.1 Comparison between Analog and Digital Transducers

Sr. No.	Analog transducers	Digital transducers
1.	Outputs of these transducers are analog in nature.	Outputs of these transducers are in the form of pulses.
2.	Convert the input quantity in analog output.	Convert the input quantity in digital output.
3.	Examples: Strain gauges, Potentiometer etc.	Examples: Rotary encoder, Digital LVDT etc.

1.38 TRANSDUCER AND INVERSE TRANSDUCER
(INPUT AND OUTPUT TRANSDUCERS)

1. **Transducer:**
   - Transducer converts a non-electrical quantity into electrical quantity.
   - **Examples:**
     (i) **Microphone:** It converts **sound signal** (non-electrical quantity) into **an electrical signal** (electrical quantity).
     (ii) **Thermistor:** It senses the **temperature** or change in temperature (non-electrical quantity) and convert it into change in **resistance** (electrical quantity).
     (iii) **Piezoelectric device:** It contains special crystals. If **pressure** (non-electrical quantity) is applied to them in one direction, the crystals will produce **voltage** (electrical quantity).

2. Inverse Transducer:

- Inverse transducer converts an electrical quantity into non-electrical quantity.
- **Examples:**
    - **(i)** **Loudspeaker:** It converts **electrical signal** (electrical quantity) in to **sound signal** (non-electrical quantity).
    - **(ii)** **Light emitting diode:** It converts **electrical signal** (electrical quantity) into **light** (non-electrical quantity).
    - **(iii)** **Piezoelectric buzzers:** It converts **electrical signal** (electrical quantity) into **sound signal** (non-electrical quantity).

1.39 ADVANTAGES AND DISADVANTAGES OF TRANSDUCERS

Advantages:

1. More compact instrumentation.
2. Good frequency and transient response.
3. Feasibility of remote indication and recording.
4. Possibility of mathematical processing of signals like summation, integration etc.
5. Minimum friction and mass inertia effect.
6. Possibility of non-contact measurement.

Disadvantages:

1. Sometimes, transducer requires external power for operating.
2. Complicated circuit.
3. High cost.
4. Transmission losses occur.

1.40 REQUIREMENTS OF A GOOD TRANSDUCER

1. Compact in size.
2. Lighter in weight.
3. Measuring system compatibility.
4. High sensitivity.
5. Ability to withstand adverse environmental conditions.
6. Low cost.
7. High range.
8. Transient response.
9. Ability to remote indication and data recording.
10. Minimum transmission losses.

1.41 SPECIFICATIONS OF TRANSDUCER

- Transducers are specified by following characteristics.
    1. Sensing system.
    2. Ranges available.
    3. Linearity and hysteresis.
    4. Output for zero input.
    5. Sensitivity.
    6. Method of cooling employed.
    7. Maximum working temperature.
    8. Variation with temperature.
    9. Temperature coefficient of zero drift.
    10. Natural frequency.
    11. Mounting details.
    12. Maximum depth.

13. Operating range.
14. Operating principle (Resistance, inductance, capacitance).
15. Excitation required.
16. Resolution.
17. Output volts (AC, DC).
18. Output power (Watts).
19. Output impedance.
20. Operating frequency range.
21. Repeatability.

1.42 CHARACTERISTICS OF TRANSDUCER OR FACTORS AFFECTING SELECTION OF TRANSDUCER FOR A GIVEN APPLICATION

- Following points should be considered, while selecting transducer for particular application:

Sr. No.	Characteristics	Details
1.	Operating range	Transducer is chosen to suit operating range with good resolution.
2.	Sensitivity	Transducer should have sufficient sensitivity for very minute or minimum changes in the input (quantity) being measured. Also, it should have minimum sensitivity to external loads.
3.	Accuracy	Transducer should have high degree of accuracy for measurement process.
4.	Frequency response	Response must be flat over desired range.
5.	Environmental compatibility	Transducer should be compatible with environment. i.e. ability to withstand against temperature and self-heating effect, magnetic fields, vibration, dust, humidity, corrosion, and supply frequency etc.
6.	Ruggedness	From mechanical and electrical point of view, transducer should be rugged.
7.	Errors	Transducer should maintain expected input-output relationship with minimum or low errors. It should have long term stability against drift.
8.	Loading effects	The transducer selected should be free from load alignment effects and temperature effects.
9.	Stability and reliability	Transducer should have high degree of stability and reliability in operating range.
10.	Static characteristics	Transducer should have low static error, linearity, low hysteresis, high resolution, high degree of accuracy and repeatability.
11.	Linearity	Linear relationship between input and output is desired.
12.	Physical parameters	Physical parameters like compact size, less weight, shape, mounting arrangement and ruggedness should be considered.
13.	Purpose	Transducer may be used for either measurement of a quantity or control the quantity or both to measure quantity and hence control.

1.43 APPLICATIONS OF TRANSDUCERS FOR MEASUREMENT OF DIFFERENT PHYSICAL QUANTITIES

Sr. No.	Physical Quantity to be measured	Suitable Transducer
1.	**Pressure**	1. Pirani gauge. 2. Piezoelectric transducer.
2.	**Temperature**	1. Resistance thermometer (RTD). 2. Thermistor. 3. Thermocouple.
3.	**Flow**	1. Hot wire anemometer. 2. Venturimeter.
4.	**Humidity**	1. Hair hygrometer.
5.	**Displacement**	1. Potentiometer device. 2. Capacitive transducer. 3. LVDT.
6.	**Force**	1. Strain gauge. 2. Load cell.

Important Points

- **Measurement** is defined as, "an act or result of comparison between a quantity (whose magnitude is unknown), with a similar quantity, such as standard (whose magnitude is known)".
- **Measurement** may also be defined as, "the process of determining the value of magnitude of an unknown quantity by comparing it with some standard or reference."
- In **Primary measurements**, required value of a parameter is determined by comparing it directly with "reference standards".
- **Secondary measurements** are defined as, "the indirect measurements involving one translation".
- **Tertiary measurements** are defined as, "the indirect measurements involving two translations or conversions".
- **International standards** are internationally agreed standards and are maintained at international bureau of weights and measures in Paris.
- **Primary standards** are maintained at national standard laboratories in different countries. They are used to calibrate and verify secondary standards.
- **Secondary standards** are maintained as reference standards in industries for the purpose of measurement and calibration of working standards.
- **Working standards** are used to check and calibrate the instruments in laboratory.
- In **Null type instruments**, deflection is made zero by applying an effect opposing to that generated by the measured quantity.
- In **Deflection type instruments**, the measurement is done by observing the relative displacement between pointer and dial.
- **Error** is the difference between measured value and true value of the measured quantity.
- **Static error** is defined as, "the difference between best measured value and true value of quantity."
- **Relative static error** is defined as, "the ratio of absolute error and measured value".
- **Errors** are **classified** into **systematic errors** and **random errors**.
- **Systematic errors** include Instrumental error, Environmental error, Operational error, Translation and signal transmission error, Observational error etc.
- **Random errors** are accidental in nature. Therefore, they cannot be traced and minimized. They are also called as residual errors or system interaction errors.
- **Static characteristics** can be defined as, "a set up criteria, which provides meaningful description of quality measurement under static conditions."
- When an instrument is used to measure an input varying with time, the behaviour of instrument is called as **dynamic or transient behaviour.**
- **Accuracy** is defined as, "the agreement of result of measured quantity with its true value".
- **Precision** is defined as, "the ability of the instrument to reproduce consistent reading for a constant input".

- **Repeatability** is defined as, "the ability of the measuring instrument to repeat the output readings consistently, when the same quantity is measured number of times under identical conditions (i.e. same observer, same instrument and same environmental conditions within short intervals of time)."
- **Reproducibility** is defined as, "the closeness of agreement between the results of measurement of the same quantity, when individual measurements are carried out by different observers, by different methods, using different instruments under different environmental conditions."
- **Reliability** is defined as, "the probability of measuring instrument, that it will perform its function for a given specific time period under given conditions."
- **Sensitivity** is defined as, "the ability of the instrument to respond to very small variation in the input (quantity being measured)."
- **Static sensitivity** of an instrument or an instrumentation system is defined as, "the ratio of magnitude of change in output signal to the magnitude of change in input signal."
- **Linearity** is defined as, "the ability of the instrument to give output, which is linearly proportional to input."
- **Non-linearity** of instrument is defined as, "deviation of calibration curve from the specified straight line."
- **Drift** is defined as, "the undesired change or a gradual variation in the output over period of time which is not related to changes in input, operating condition or load."
- **Hysteresis** is defined as, "the non-coincidence of loading and unloading curves." Hysteresis is determined by the maximum difference in any part of output readings.
- **Backlash** is defined as, "the maximum distance or angle through which any part of mechanical system may be mover in one direction without applying force or motion to the next part in a mechanical system".
- **Dead zone** is defined as, "the largest range through which, an input signal can be varied without any change in output of instruments."
- **Dead time** is defined as, "the time required for measurement system to begin to respond to change in the measurand."
- **Threshold of instrument** is defined as, "the minimum value, below which, no output voltage change can be detected, when input of an instrument is increased gradually from zero."
- **Threshold** can also be defined as, "the minimum value of input, which is necessary to cause a detectable change from zero output."
- **Resolution** is defined as, "the small increment in the input quantity from a non-zero value being measured, which can be detected by an instrument."
- **Resolution** can also be defined as, "the smallest measurable change in input, for which, there will be a change in output."
- **Speed of response** is defined as, "the value of rapidity (rapidness) with which, an instrument responds to the change in the quantity being measured".
- **Measuring lag** is defined as, "the retardation or delay in the response of an instrument to measure a given quantity."
- **Fidelity** is defined as, "the degree, to which, a measurement system indicates change in the measured quantity without any dynamic error."
- **Fidelity** is also defined as, "the degree of closeness, with which, the system indicates or records the signal, which is impressed upon it."
- **Dynamic error** is defined as, "the difference between the true value of quantity (under measurement), changing with time and the value indicated by measurement system."
- **Overshoot** is defined as, "the maximum amount, at which, the pointer moves beyond the steady state."
- **Calibration** is defined as, "the process of framing of scale of instrument by applying some standard."
- **Range** is defined as, "the region between the limits, within which, an instrument is desired to operate for measuring, indicating or recording a physical quantity."
- **Span** is defined as, "the algebraic difference between upper and lower measurement limits of any instrument."
- **Transducer** is a device, which senses the measurand (quantity to be measured) as input signal and converts it into an output signal to indicate measured value.
- **Primary sensing element (Detector)** is the part of transducer, which responds to a physical phenomenon or change in the physical phenomenon.
- **Transduction element** transforms the output of primary sensing element into some another signal. It is also referred as secondary transducer.
- **Mechanical transducer** converts a physical quantity or mechanical parameter into an equivalent output of some other form, which is the measure of quantity under measurement.

- **Electrical transducer** converts a physical quantity or mechanical parameter into an equivalent electric signal.
- In **Primary transducer**, there is only one stage of transduction. Primary transducer makes the first contact with the measurand, senses it as a physical parameter (like temperature, pressure) and converts it into suitable and readable physical parameter (like displacement).
- In **Secondary transducer**, there are two stages of transduction. Here, the primary sensor of the instrument makes the first contact with the quantity to be measured and then converts it into some other suitable form of signal, which represents the measured value.
- **Active transducers** do not require external power to operate. The energy required to generate output signal is obtained from the physical quantity being measured. Therefore, they are known as self-generating transducers.
- **Passive transducers** require an external source of power for their operation. Therefore, they are known as power operated transducer.
- **Analog transducer** converts the input quantity into an analog output, which is a continuous function of time.
- **Digital transducer** converts the input quantity into an electrical output, which is in the form of pulses or digitally coded form.
- **Transducer** converts a non-electrical quantity into electrical quantity.
- **Inverse transducer** converts an electrical quantity into non-electrical quantity.

Practice Questions

1. Define measurement. Give classification of measurement.
2. What is measurement? State its basic requirement and significance.
3. List desirable and undesirable static characteristics of an instrument.
4. Explain the difference between accuracy and precision in an instrument.
5. Differentiate between accuracy and precision with suitable example.
6. Define sensitivity drift and zero drift. What factors can cause sensitivity drift and zero drift in instrument characteristics?
7. State and explain any two dynamic properties of measurement system.
8. Define error. Explain the detail classification of error.
9. Explain the difference between systematic and random errors. What are the typical sources of these two types of errors?
10. What are different types of errors in measurement system? Give classification.
11. Define the terms: Threshold and Resolution.
12. Explain overshoot in a measuring instrument.
13. Define the terms: Range and Span.
14. Define: (i) Speed of response, (ii) Drift, (iii) Dead zone, (iv) Span.
15. What is drift? Explain with sketch.
16. What is systematic error in measurement? How it can be reduced?
17. What is dead zone? What are the factors responsible for dead zone?
18. State at least four important characteristics of the measuring instruments.
19. Define "Span" of an instrument. How is it different from "range"?
20. What is indirect method of measurement? Give any one example.
21. Define overshoot with a sketch.
22. Explain the importance of calibration of instrument during measurement.
23. Explain the need of calibration of measuring instruments.
24. State any two considerations at the time of instrument calibration.
25. Define static characteristic of a measuring instrument. State any four static characteristics of measuring instruments.
26. Define: (i) Calibration, (ii) Precision, (iii) Hysteresis, (iv) Drift.
27. Differentiate between repeatability and reproducibility.
28. What is function of transducer? Differentiate between active and passive transducer.
29. Define transducer. Explain the classification of transducers with suitable example.
30. Define transducer and give its classifications.
31. Classify transducers and explain any one with diagram.
32. How transducers are specified? Explain.
33. Compare between active and passive transducers (Four points).

■■■

Chapter 2

DISPLACEMENT AND TEMPERATURE MEASUREMENTS

2.1 CAPACITIVE TRANSDUCER

Principle of Working:

- Capacitive transducer works on the **principle of change in capacitance.**
- Capacitance changes, when there is,
  (i)   Change in overlapping area 'A' of plates.
  (ii) Change in distance 'd' between two plates.
  (iii) Change in dielectric constant 'K'.

Construction:

- Capacitive transducer consists of two or more metal plate conductors separated by a dielectric medium or material, which acts as an insulator.
- The dielectric material may be air, mica, oil, paper etc. Fig. 2.1 shows capacitive transducer, where air is dielectric medium existing between two parallel metal plates.
- Value of dielectric constant (K) is different for different dielectric medium or material.

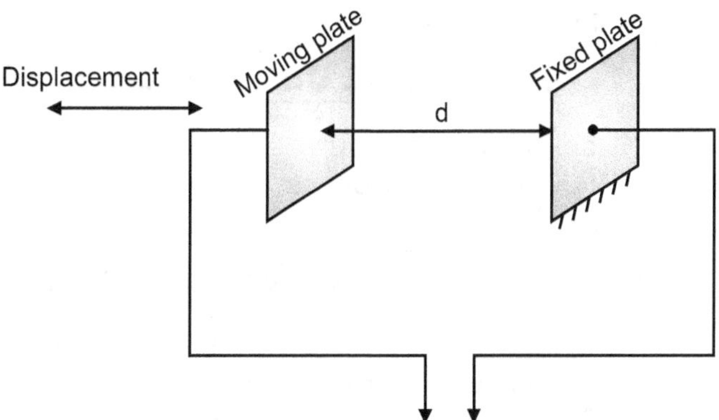

Fig. 2.1: Capacitive Transducer

Working:

- When voltage is applied across the plates, equal and opposite charges are generated on the plates.
- The capacitance of a parallel plate capacitor is given by the equation,

$$C = \frac{KA}{d} \qquad \qquad \text{... (2.1)}$$

where,   A = Area of each plate or Overlapping area of plates in 'm^2'

         d = Distance between two plates   in 'm'

         K = Dielectric constant of medium existing between parallel plates

- Thus, the capacitance 'C' can be changed by changing the di-electric medium (i.e. dielectric constant "K") or changing the area of plates 'A' or changing distance 'd' between plates. This results into change in output voltage of circuitry.
- These changes can be caused by physical variables such as displacement, force, pressure.
- From Fig. 2.1, it can be observed that, one plate is fixed, whereas, second plate is capable to move. Also, it can be seen that, plates having same cross-sectional area are placed or located in such a way that, overlapping area of plates (A) will remain unchanged, if moving plate is moved towards or away from the fixed plate in horizontal direction. Similarly, dielectric medium, i.e. air will remain unchanged during horizontal displacement of moving

plate. Therefore, value of dielectric constant (K) will be constant. Therefore, equation (2.1) can be written as,

$$C \propto \frac{1}{d} \qquad (\because \text{'A' and 'K' are constant})$$

- When moving plate is displaced horizontally leftwards or rightwards, value of capacitance (C) will change due to change in distance 'd' between the two plates, which can measured.
- For measurement, calibrated voltmeter is connected to output terminals of circuitry, which can be calibrated to give a direct reading of change in distance between two plates, i.e. displacement of moving plate.
- Change in capacitance can also be measured by an A.C. bridge, which gives the change in distance between two plates. Hence, linear displacement can be measured.
- Capacitive transducers are also used for measurement of angular displacement.

2.1.1 Advantages of Capacitive Transducer

1. It requires very small force to operate, hence it can be used in small system also.
2. Excellent frequency response.
3. Extremely sensitive.
4. Easy to fabricate.
5. Require small power to operate.
6. Simple in construction.

2.1.2 Disadvantages of Capacitive Transducer

1. Non-linear behaviour.
2. High output impedance.
3. Poor frequency response.

2.1.3 Applications of Capacitive Transducer

1. For measurement of both linear and angular displacements.
2. For measurement of humidity in gases.
3. For measurement of volume, density, liquid level, weight etc.

2.2 LINEAR POTENTIOMETER

- This is a device, which converts mechanical displacement (to be measured) into an electrical output.
- Linear potentiometer can measure displacements in the range of 2.5 mm to 500 mm.

Principle of Working:

"Change in the position of slider leads to change in resistance of potentiometer wire and corresponding change in output voltage generated is a measure of displacement of slider to be measured".

Construction:

- Linear potentiometer consists of a stretched resistance wire and a sliding or movable contact (wiper).
- The resistance element or resistance wire is made up of metal alloy. This resistance wire is wound on a former in such a way that, the slider or wiper may be moved along the various turns axially (i.e. translational or rectilinear movement).
- Former is a cylindrical piece made up of an insulating material, such as, ceramic or plastic. Former is also known as **mandrel**.
- The slider or wiper always maintains an electrical contact with the resistance wire and moves over it. The component, whose displacement is to be measured, is attached to slider or wiper. Linear potentiometer is also called as 'Linear POT'.

Working:

- Linear potentiometer is a passive transducer, because it requires external power source for its operation.

- Therefore, the resistance wire is excited with either AC or DC voltage. It is represented as input voltage (e_i).

- When the slider moves or slides axially along the various turns of resistance wire, the effective resistance existing between one end of wire and slider also changes.

- Due to this, an output voltage (e_o) is generated, which can be measured.

- Alternatively, this output voltage generated can be directly calibrated to give displacement.

- The output voltage generated is a linear function of the displacement to be measured.

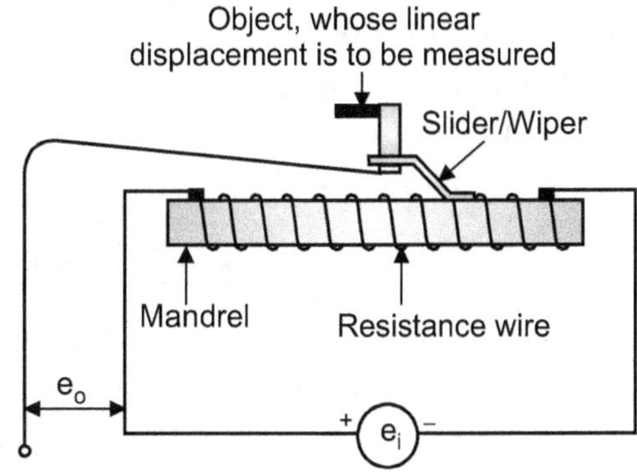

Fig. 2.2: Linear Potentiometer

- Linear displacement of slider and hence, the component is calculated by using the following mathematical relation,

$$x_i = x_t \times \left(\frac{e_o}{e_i}\right)$$

where,

e_i = Input voltage

e_o = Output voltage

x_t = Total length of linear pot (i.e. translational potentiometer)

x_i = Displacement of slider from its zero position

R_p = Total resistance of potentiometer in Ω (ohms)

- Above mathematical relation of x_i can be derived in the following manner.

- If the variation of resistance with respect to displacement is linear, then resistance per unit length is given by $\left(\frac{R_p}{x_t}\right)$.

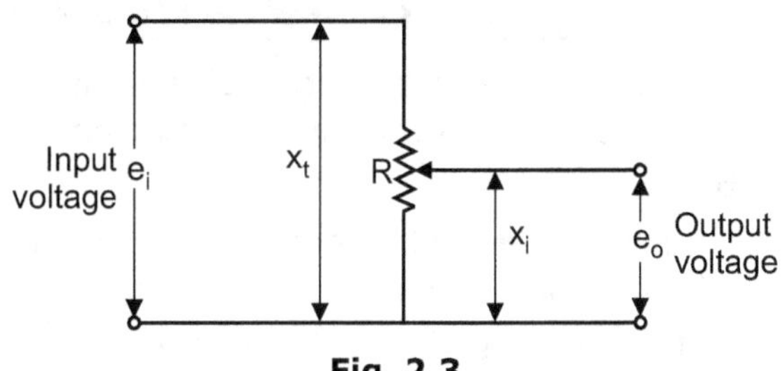

Fig. 2.3

- The output voltage under ideal condition is given by,

$$e_o = \left(\frac{\text{Resistance at the output terminals}}{\text{Resistance at the input terminals}}\right) \times \text{Input voltage} = \frac{\frac{R_p}{x_t} \times x_i}{R_p} \times e_i$$

$$\therefore \quad e_o = \frac{x_i}{x_t} \times e_i$$

$$\therefore \quad \frac{e_o}{e_i} = \frac{x_i}{x_t}$$

$$\therefore \quad x_i = x_t \times \left(\frac{e_o}{e_i}\right)$$

- Under ideal circumstances, the output voltage varies linearly with displacement.

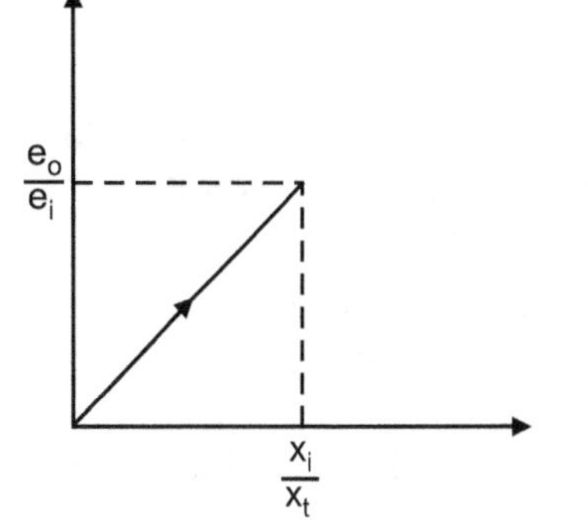

Fig. 2.4: Input-Output Relationship of Potentiometer

2.2.1 Advantages of Potentiometer

1. Low cost.
2. Simple to operate.
3. Useful for measurement of large amplitudes of displacement.

4. Electrical efficiency is very high and provides sufficient output to allow control operations.
5. Rugged construction.
6. Low maintenance.
7. Less error in measured value, since reading is not affected due to vibration and changes in temperature.
8. Suitability for A.C. excitation method.

2.2.2 Disadvantages of Potentiometer

1. For linear potentiometer, a large force is required to move the slider. So, it is not suitable for small force actuation.
2. If the sliding contact becomes misaligned, then operation becomes noisy.
3. Resolution is poor.
4. Poor dynamic response due to friction involved.
5. The device has limited life due to early wear of wiper or slider.

2.2.3 Applications of Potentiometer

1. For measurement of pressure.
2. For measurement of linear and angular displacement.
3. Both linear and rotary type potentiometers are used in controlling audio equipments for changing loudness, frequency of audio signals etc.
4. Used as position feedback devices in closed loop control systems, such as, servomechanism.

2.3 ROTATIONAL TYPE POTENTIOMETER

- Rotational type potentiometer is used for measurement for angular displacement.

Construction:

- Resistive elements are circular in shape in case of rotational type potentiometer.
- The resistive elements of potentiometer are metal alloys wound on a cylindrical ceramic former.
- Wiper (slider) is made up of phosphor bronze and it is gold plated for good electrical contact.

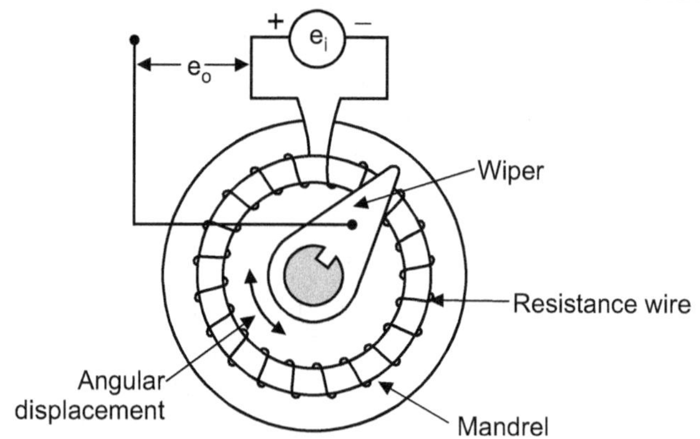

Fig. 2.5: Rotary Potentiometer

Working:

- The resistance wire is wrapped around a circular former, so that, wiper may be rotated along the various turns along an arc (angular movement). The circular former is also called as mandrel.
- The wiper always maintains an electrical contact with the resistance wire wound on a circular former made up of plastic or ceramic (i.e. any insulating material) and rotates over it.
- The resistance element is excited with either AC or DC voltage.
- The output voltage generated is a linear function of angular displacement (θ_i) to be measured.

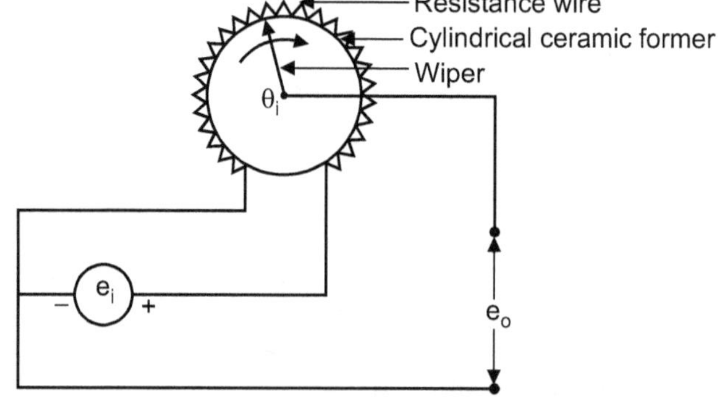

Fig. 2.6

2.4 MATERIALS USED FOR POTENTIOMETER

1. Resistance Wire:
- Commonly used material for resistance wire is **alloy metal**.
- However, depending upon the specific requirements and applications, it is also made up of **platinum, chromium, nickel, copper** etc.
- Resistance wire has to carry large current at high temperature. Therefore, temperature coefficient of resistance wire material should be small.

2. Wiper or Slider:
- Wiper (slider) is made up of **phosphor bronze**.
- For good electrical contact, it is gold plated.

2.5 LINEAR VARIABLE DIFFERENTIAL TRANASFORMER (LVDT)

- LVDT is the most commonly used inductive transducer to translate linear motion into electrical signal (displacement).
- Here,   L = Linear motion (displacement)

  V = Variable inductance

  D = Differential, which means output is difference of two secondary outputs

  T = Transformer, because it functions as the former of primary and secondary windings.

Working Principle:
- *Position of iron core (magnetic core or armature) is responsible for varying the differential voltage of two secondary windings of a transformer, which is a measure of linear displacement.*

Construction:
- LVDT consists of an insulating hollow cylinder made up of Bakelite or any insulating material.
- On the circumference of this insulting cylinder, one primary winding (P) and two secondary windings (S_1 and S_2) are wound.
- The primary winding (P) is wound at the centre of insulating cylinder and on either side of it, two secondary windings (S_1 and S_2) are wound exactly opposite to each other. Thus, the directions of S_1 and S_2 are exactly opposite to each other.
- Inside the hollow insulating cylinder, magnetic or armature core is placed, which is free to move back and forth.
- The displacement to be measured is attached to the arm of soft iron core/magnetic core.
- The magnetic core is made up of nickel iron, which gives high sensitivity and low null voltage. This core is slotted longitudinally to reduce eddy current losses.
- The whole assembly is placed in stainless steel housing.

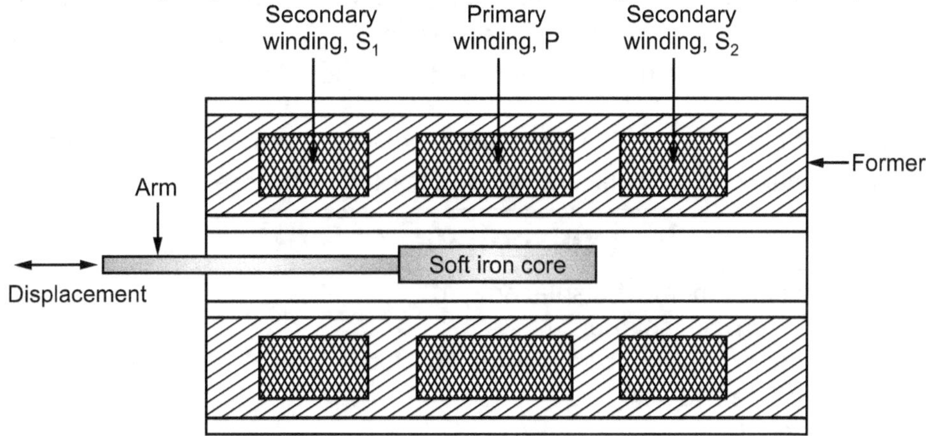

Fig. 2.7: LVDT

Working:

- When A.C. supply is given to the primary winding, it produces magnetic flux, which completes its path through S_1 and S_2.
- While completing the path, the flux produced by the primary winding is linked to the number of conductors of secondary windings.
- Therefore, according to Faraday's law of electromagnetic induction, an e.m.f. is produced in the secondary winding.
- As shown in Fig. 2.8 (a), the output voltage of S_1 is E_{s_1} and that of S_2 is E_{s_2}.
- To convert the output of S_1 and S_2 into single voltage signal, S_1 and S_2 are connected in series as shown in Fig. 2.8 (b), with phase difference of 180° (wounded in opposite to each other).
- Thus, the output of transducer will be the difference of two voltages i.e. differential output voltage. Therefore, $E_o = E_{s_1} - E_{s_2}$

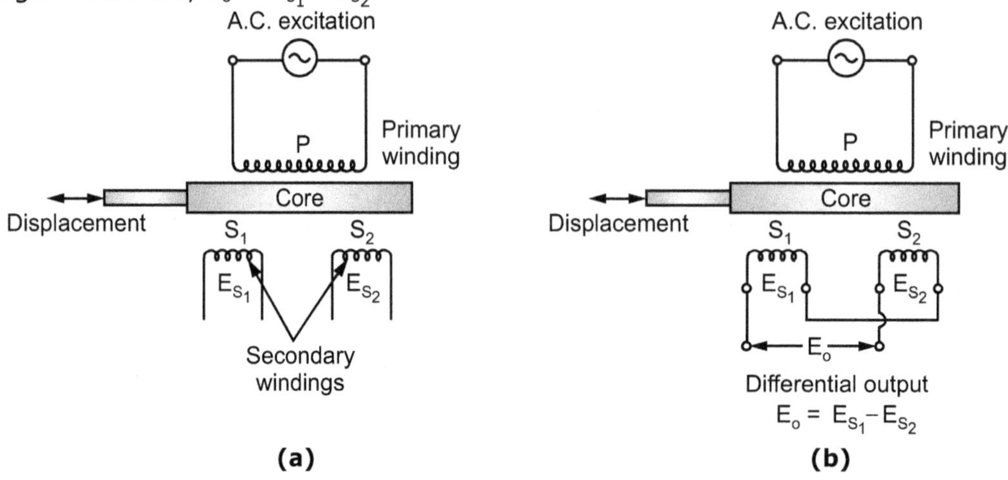

Fig. 2.8: Electrical Circuit of LVDT

Various Positions of Core and Corresponding Displacement Measurements:

1. When the core is at **central** position of insulting cylinder, the magnetic flux linking to both S_1 and S_2 is equal and hence, equal e.m.f.s. are induced in S_1 and S_2.

   Thus, $\qquad\qquad E_{s_1} = E_{s_2}$

   Therefore, $\qquad\qquad E_o = E_{s_1} - E_{s_2} = 0$

   This position of core is called as normal (null) or zero position. It concludes that, output voltage is zero at null position.

2. When core is moved towards left, then flux linking to coil S_1 increases and at the same time, flux linking to coil S_2 decreases. Hence, e.m.f. induced in coil S_1 (E_{s_1}) is greater than of S_2 (E_{s_2}). As a result of which, output measurement between secondary terminals is in phase with primary voltage.

3. When core is moved towards the right side, then flux linking to coil S_2 (E_{s_2}) is more as compared to coil S_1 (E_{s_1}). Thus, e.m.f. induced in S_2 (E_{s_2}) is greater than e.m.f. in S_1 (E_{s_1}). As a result of which, output measurement between secondary terminals is out of phase with primary voltage.

- Hence, output voltage ($E_{s_1} - E_{s_2}$) measured between two secondary terminals of LVDT is directly proportional to the displacement of soft magnetic core.

2.5.1 Advantages of LVDT

1. High range for measurement of displacement.
2. Less friction and electrical isolation.
3. Immunity (protection or exemption from obligation/penalty) for external effects.
4. High input and high sensitivity.
5. Rigid in construction.
6. Low hysteresis. i.e. very small amount of current remains in electrical circuit.

7. Low power consumption.
8. Linearity i.e. Linear output over a wide range of motion.
9. High output voltage.
10. Infinite resolution.

2.5.2 Disadvantages of LVDT

1. Large displacement requirement.
2. The frequency of carrier must be at least ten times the highest frequency of component to be measured.
3. Temperature affects the performance of transducer.
4. Vibrations affect the performance of transducer.
5. Residual voltage problem.
6. Limited dynamic response.

2.5.3 Applications of LVDT

1. LVDT can be used for displacement measurement ranging from fraction of 1 mm to a few cm.
2. Acting as a secondary transducer, LVDT is used to measure force, weight and pressure.
3. Useful for measurement of tension in rod, cord, cable, wire rope etc.
4. Useful for measurement and control of thickness of metal sheet.
5. Useful for measurement of displacement in CNC machines.

2.6 CHARACTERISTICS OF LVDT

1. The output voltage is proportional to the displacement of iron core.
2. Negligible operating force and no wear of moving parts.
3. The device is very sensitive and it is capable to sense direction of movement of displacement sensor and accordingly, it gives ($\pm$ x) output.
4. Simplicity of design, ease of fabrication and installation, rugged and durable construction.
5. It is insensitive to the temperature change of the measurement system.
6. High output, low hysteresis, continuous (infinite) resolution and linear electrical response.
7. Small physical size and better repeatability.
8. Ideally, the output voltage at null position should be zero, but in actual practice, there exists a small voltage at null position. This is because of presence of some harmonic in the input supply voltage due to the use of iron core. Due to this, incomplete magnetic or electrical unbalance results into a finite output voltage at null position. This finite output voltage at zero displacement is also called as **residual voltage**. This is less than 1% of maximum output voltage. With improved technological methods and with use of better a.c. sources, residual voltage can be reduced up to negligible value.
9. When the core is moved in one direction from the null position, the differential output voltage will increase. This differential voltage i.e. output voltage of LVDT is a linear function of core displacement within a limited range of movement; say about 5 mm from null position. **The figure shows variation in output voltage against displacement.** For different positions of core, the curve is practically linear for small displacement (say about 5 mm).

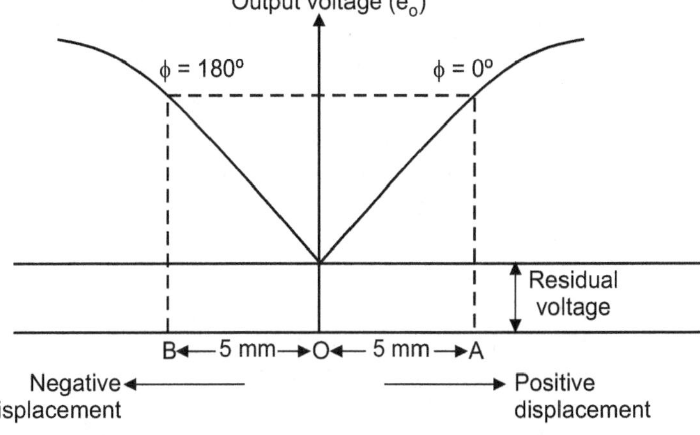

Fig. 2.9: Displacement versus Output Voltage

Beyond this range of displacement, the curve starts to deviate from straight line.

2.7 SPECIFICATIONS OF LVDT

(i)   Range = 0 to 50 mm.
(ii) Accuracy = 0.1 % of range
(iii) Ambient temperature = – 40° to + 50° C.
(iv) Input frequency = 20 to 50 kHz.
(v)   Resolution = 2×10^{-3} mm
(vi) Input or excitation voltage = 3 to 15 V (sinusoidal)
(vii) Stroke or core displacement = 0.1 mm to 100 mm.

2.8 ROTARY VARIABLE DIFFERENTIAL TRANSFORMER (RVDT)

- RVDT is used to sense the angular displacement.
- It is an inductive transducer, which converts angular displacement into an electrical signal.
- Here,   R = Rotary motion / Angular displacement
             V = Variable inductance
             D = Differential, which means output is difference of two secondary outputs
             T = Transformer, because it functions as the former of primary and secondary windings.

Working Principle:

Rotation of iron core (magnetic core/armature) is responsible for varying the differential voltage of two secondary windings of a transformer, which is a measure of angular displacement.

Construction:

- RVDT consists of two secondary windings having equal number of turns, but wound in opposite direction of primary winding connected across single phase A.C. supply.
- Magnetic core or armature is placed in the air gap between primary and secondary windings.
- It is free to move in either direction depending on angular displacement of shaft.

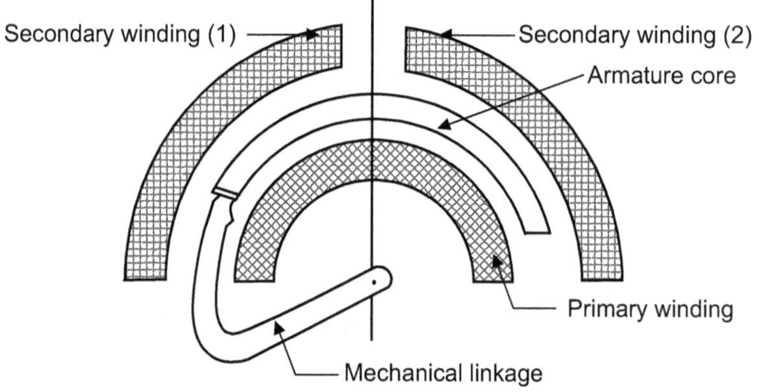

Fig. 2.10: Illustrative View of RVDT

Working:

- When single phase A.C. supply is given to the circuit, it produces magnetic flux, which completes its path through S_1 and S_2.
- While completing the path, the flux produced by primary winding is linked to number of conductors of secondary windings.
- Therefore, according to faraday's law of emf, an emf is produced in secondary winding.
- As shown in Fig. 2.11, the output voltage of S_1 is E_{s_1} and that of S_2 is E_{s_2}.
- To convert the output of S_1 and S_2 into single voltage signal, S_1 and S_2 are connected in series as shown in Fig. 2.11.
- Thus, the output of transducer will be the difference of two voltages i.e. differential output voltage. i.e. $E_o = E_{s_1} - E_{s_2}$

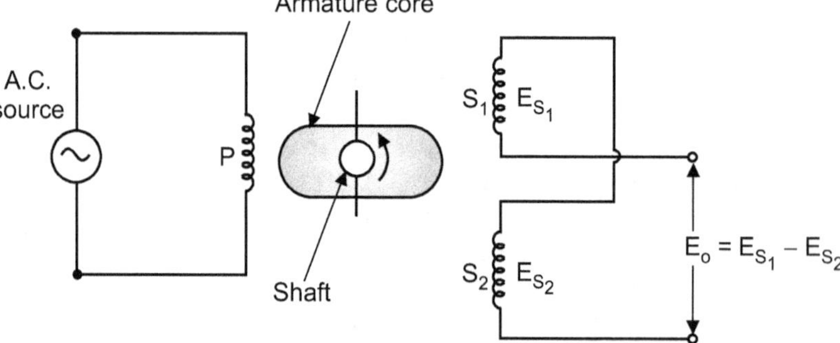

Fig. 2.11: RVDT

Various Positions of Core and Corresponding Displacement Measurements:

1. When the core is at central position, the magnetic flux linking to both S_1 and S_2 is equal and hence, equal emfs are induced in S_1 and S_2. Thus, $E_{s_1} = E_{s_2}$.

   Therefore, $\quad\quad\quad E_o = E_{s_1} - E_{s_2} = E_{s_1} - E_{s_1} = 0$

   This position of core is called as normal (null) or zero position.

   It concludes that, output voltage is zero at null position.

2. When the core rotates anti-clockwise, then the flux linking to coil S_1 increases and at the same time, the flux linking to coil S_2 decreases. Hence, emf induced in coil S_1 (E_{s_1}) is greater than that of S_2 (E_{s_2}). As a result of which, output measured between secondary terminals is in phase with primary voltage.

3. When the core rotates clockwise, then the flux linking to coil S_2 increases and at the same time, the flux linking to coil S_1 decreases. Hence, emf induced in coil S_2 (E_{s_2}) is greater than that of S_1 (E_{s_1}). As a result of which, output measured between secondary terminals is out of phase with primary voltage.

   Hence, output ($E_{s_1} - E_{s_2}$) measured between two secondary terminals of RVDT is directly proportional to angular displacement of magnetic core.

2.8.1 Advantages of RVDT

1. Infinite resolution and linear operation.

2. Ability to measure clockwise as well as anti-clockwise angular displacement.

2.8.2 Disadvantages of RVDT

1. Range of RVDT is limited up to $\pm 40°$ only, even though, it is a 360° device.

2. Better linearity is obtained only, if small angular displacements within $\pm 5°$ are measured.

2.8.3 Applications of RVDT

1. In Industrial applications.

2. For steering, braking operations in automotive applications.

3. Defense applications, where low noise is expected arising due to friction.

4. Flight control applications.

2.9 CHARACTERISTICS/ SPECIFICATIONS OF RVDT

1. RVDT is a continuous rotational device (360°).

2. High resolution to very small fraction of degree can be achieved in practice by RVDT.

3. The range of most linear operations for a typical RVDT is only about $\pm 40°$.

4. The range of a typical RVDT is only $\pm 40°$ as compared to theoretically described 360° rotation phenomenon. However, if RVDT is used to measure angular displacement within $\pm 5°$, we get an improved accuracy.

5. The foremost advantage of RVDT is the lack of physical contact between rotor and stator.

6. RVDT shows linearity over smaller angular displacement.

7. The resolution of RVDT is theoretically definite. Resolution of small fractions of degree is generally achieved in practice.

2.10 SPECIFICATION OF DISPLACEMENT TRANSDUCER

- Sample specification of a displacement transducer is given below.

Specification	Capacity	Non-linearity	Rated output	Any other
Sliding resistance wire	30 mm....300 mm	0.2	1.5 ± 1%	
LVDT	0 to 50 mm	Linearity over small displacement		Accuracy 0.1 % of range
RVDT	40°	Linearity over small displacement		Accuracy 0.1 % of range
Dial gauge type	5 mm.....30 mm	0.5	0.5, 1, 1.5, ±20%	

2.11 SELECTION OF DISPLACEMENT TRANSDUCER

- Following factors should be considered, when selecting a displacement transducer.
    1. Required accuracy.
    2. Resolution required.
    3. Size of displacement.
    4. Type of displacement.
    5. Cost of transducer.
    6. Material used for transducer.

2.12 APPLICATIONS OF DISPLACEMENT TRANSDUCER

- Applications of displacement transducers such as Potentiometer, LVDT, RVDT etc. are given below.
    1. LVDT can be used in all displacement measurements, where displacement ranges from fraction of mm to few centimetres.
    2. LVDT is used in CNC machines for displacement measurement.
    3. LVDT can be used for measurement and control of thickness of a metal sheet being rolled.
    4. Used for measurement of tension in channel.
    5. RVDT is used in flight control system.
    6. Potentiometer is used in design of many transducers for measurement of pressure, force and acceleration.
    7. Potentiometer can be used as a position feedback in servomechanism.
    8. Acting as a secondary transducer, displacement transducer can be used as a device to measure force, weight, pressure etc.

2.13 TEMPEATURE MEASUREMENT

- **Temperature** is defined as, **"the degree of hotness and coldness of a body or an environment measured on definite scale."**
- Also, **Temperature** can be defined as, *"the driving force or potential causing flow of energy in the form of heat."*
- Temperature can be measured in various units, such as, Degree Fahrenheit, Degree Celsius, Kelvin and Rankine.

2.14 CLASSIFICATION OF TEMPERATURE MEASURING INSTRUMENTS

- **Temperature Measuring Instruments may be classified,**
  1. According to the range of temperature measurement, or
  2. According to nature of change produced in the temperature sensing element.

Classification of Temperature Measuring Instruments

Mechanical Thermometers	Electrical Thermometers	Radiation Pyrometers

1. Metallic expansion thermometers
  (a) Expansion rod thermometer
  (b) Bimetallic thermometer
2. Filled system thermometers
  (a) Liquid filled thermometer
  (b) Mercury filled thermometer
  (c) Vapour pressure thermometer
	1. Metallic resistance thermometer (RTD)
	2. Thermistor
	3. Thermocouple
		1. Total radiation pyrometer
		2. Optical pyrometer

2.15 NON-ELECTRICAL METHODS OF TEMPERATURE MEASUREMENT

- Following are the instruments working on Non-electrical methods of measurement:
  1. Bimetallic thermometers.
  2. Liquid in glass thermometer (Pressure thermometer).
  3. Liquid in metal thermometer (Pressure thermometer).
  4. Vapour pressure thermometer.

2.16 WORKING PRINCIPLE OF BIMETALLIC THERMOMETERS

- Bimetallic thermometers use **two fundamental principles**:
  (a) All metals expand or contract with change in temperature.
  (b) Temperature coefficients of expansion are not same for all metals and therefore, their rates of expansion or contraction are different. This property of difference in rates of thermal expansion is used to produce deflections proportional to temperature changes.
- **A bimetallic thermometer** consists of a bimetallic strip, which is constructed by bonding together two thin strips of two different metals, such that, they cannot move relative to each other.
- Bimetallic strips can be arranged in various forms like flat, spiral and single helix and multiple helix configurations.
- As the temperature applied to the strip increases, there will be deflection of free end of strip. Length of both metals will change according to their individual coefficient of thermal expansion. Let 'α' be the notation used for **coefficient of thermal expansion.**
- As one end of bimetallic strip is fixed, the strip will bend at free end, but towards the side of metal having low coefficient of thermal expansion.
- Reverse will happen, if there is decrease in temperature.
- Deflection of free end is directly proportional to two quantities,
  1. Square of length of metal strip and,
  2. Change in temperature.
- Fig. 2.12 shows a bimetallic strip in the form of a straight cantilever beam. With one end fixed, the temperature change causes the free end to deflect.

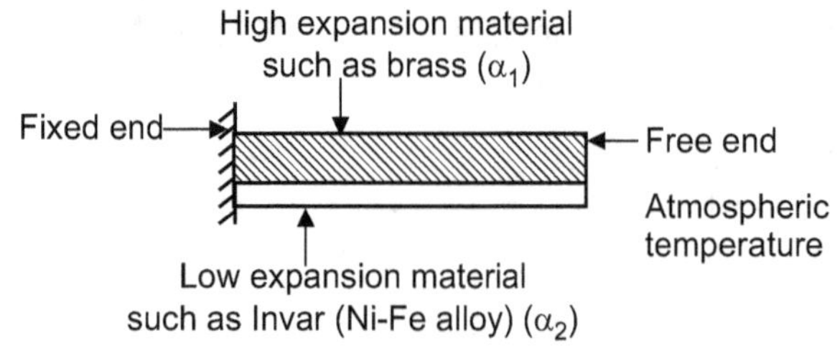

(a) Atmospheric condition

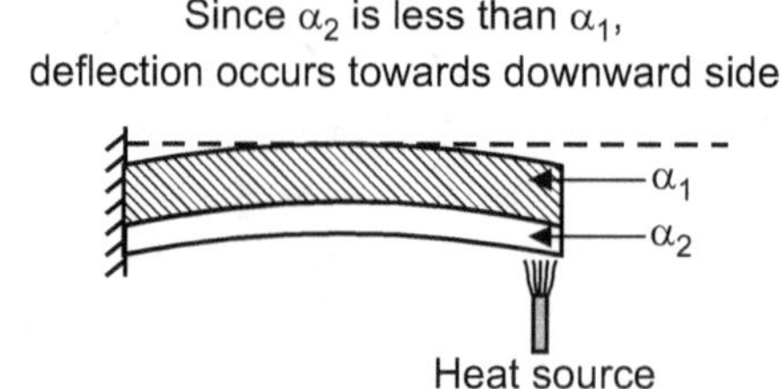

(b) A heat source is placed at free end

Fig. 2.12: Plain Bimetallic Thermometer

2.16.1 Factors Affecting the Response of Temperature Sensing Element of Bimetallic Thermometer

1. Values of heat transfer coefficient.

2. Thermal conductivity of temperature sensing element, whether low or high.

3. Change in physical dimensions due to change in temperature may have adverse effect on measured reading and accuracy of measuring instrument.

4. Surface area per unit mass of temperature sensing element.

2.17 HELIX TYPE BIMETALLIC THERMOMETER

Operating Range: 30°C to 300°C

- Fig. 2.13 shows **Helix type bimetallic thermometer**, in which, a bimetallic strip is wound in the form of helix.

- One end of helix is fastened to the casing of bulb and another end is connected to pointer.

- When the bimetallic helix is heated, difference in thermal expansion of metals causes it to unwound.

- Due to this, pointer moves or sweeps over a circular scale to indicate the measured value of temperature.

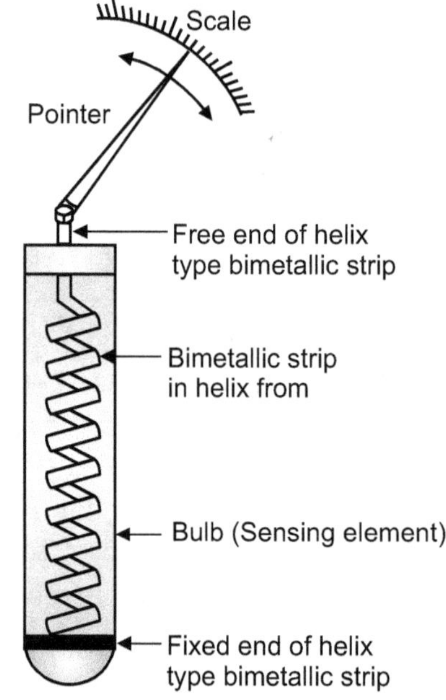

Fig 2.13: Helix Type Bimetallic Thermometer

2.17.1 Applications of Helix Type Bimetallic Thermometer

1. Used in industries to measure temperature.
2. Used in refineries, hot tanks, hot water heaters etc.

2.17.2 Advantages of Helix or Spiral Type Bimetallic Thermometer over Plain Bimetallic Thermometer

- **Plain Bimetallic Thermometer** is unsuitable for use in industrial applications due to following reasons,
  1. Less deflection of pointer, and
  2. Low sensitivity.

- In case of Plain Bimetallic Thermometer, deflection of pointer is very small and limited due to small length of bimetallic strip used. This is because, deflection of free end is directly proportional to square of length of metal strip. i.e. $\delta \propto L^2$.

- Therefore, to have more deflection and sensitivity to measure high temperatures, length of bimetallic strip needs to be increased considerably.

- But, such a longer plain bimetallic thermometer will occupy more space and its practical use is not feasible. This difficulty is overcome in case of helix or spiral type bimetallic thermometers.

- Sensitivity can be increased by increasing the length of strip. Therefore, if required length of strip is too long, then it can be shaped to a spiral or helix form, so that, the instrument will be compact in size.

- This arrangement permits a very long length of strip to be contained into a fairly small space.

2.18 SPIRAL BIMETALLIC THERMOMETER

Operating Range: 30°C to 550°C

- **Fig. 2.14** shows **a spiral bimetallic thermometer instrument**. Here, the spiral bimetallic (strip) element is attached to the pointer mounted in housing with a scale.

- When the spiral bimetal is heated, it bends in the direction of metal having low thermal coefficient of expansion, and rotates in clockwise direction. Thus, the pointer attached to it also moves on calibrated scale, indicating the temperature reading.

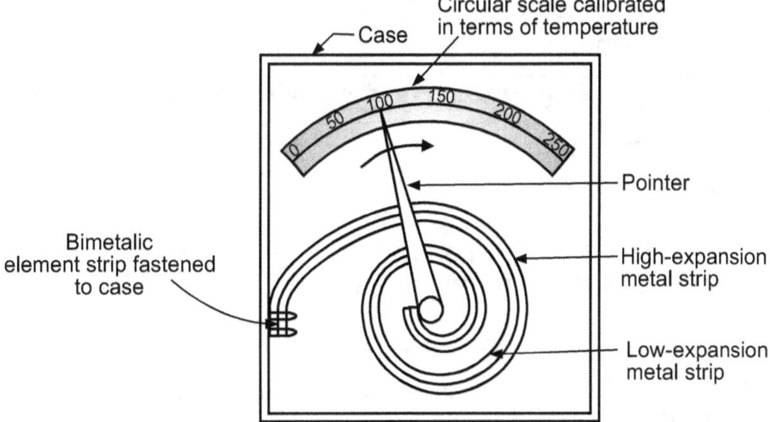

Fig. 2.14: Spiral Bimetallic Thermometer Instrument

2.18.1 Applications of Spiral Bimetallic Thermometer Instrument

1. Homes and offices for measuring ambient air temperature.
2. Commercial and Industrial applications.

2.19 ADVANTAGES, DISADVANTAGES AND APPLICATIONS OF BIMETALLIC THERMOMETERS

2.19.1 Advantages of Bimetallic Thermometers

1. Low cost.
2. Easy to install and maintain.
3. Good accuracy up to $\pm$ 0.5 %.
4. Wide temperature range.
5. Ability to withstand, if subjected to a higher temperature, which is 1.5 times more than the design temperature, for which, it is manufactured.
6. Not easily broken.
7. Simple and compact instrument.
8. Robust construction.

2.19.2 Disadvantages of Bimetallic Thermometers

1. Bimetallic thermometers do not have ability to measure temperatures, which are rapidly changing. Therefore, they are not recommended to use for measurements of fast changing temperatures.
2. Materials of bimetallic thermometers may be subjected to permanent deformation or distortion, due to which, they may not be restored to original dimensions. Under such conditions, temperatures measured and indicated may be wrong or incorrect.
3. Low speed of response.

2.19.3 Applications of Bimetallic Thermometers

1. ON or OFF type thermostats.
2. Temperature control systems.
3. Control switches in domestic ovens, electric irons, refrigerator etc.
4. Industrial applications, such as, refineries, oil burners etc.
5. Lamp flashers.
6. Time delay circuits.

2.20 PRESSURE THERMOMETER

- There are two types of pressure thermometers:
    (a) Liquid pressure thermometer.
    (b) Vapour pressure thermometer.

2.21 LIQUID IN GLASS THERMOMETER (LIQUID PRESSURE)

- **Operating Range:** Up to 300°C.
- **Method used:** Expansion of Liquids.

Working principle:

- *If temperature increases, liquid expands in volume. This change in volume of liquid is directly proportional to change in temperature.*

Construction:

- **Liquid in glass type thermometer** is a thick-walled glass tube casing.
- Another tube of very small diameter is enclosed inside this glass tube casing. This tube is called as **capillary tube.**
- Lower end of capillary tube is shaped in the form of spherical or cylindrical bulb, which is filled with responsive liquid. Generally, **Mercury** is used as responsive or thermometric liquid. Capillary tube is sealed at top.
- Accidently, if thermometer comes in contact of extreme high temperature, which is more than its maximum temperature range, then the liquid inside the capillary tube will expand in volume to a much greater extent. It can lead to breakage or damage of capillary tube and thermometer will no longer be able to measure temperature.
- To avoid such circumstances, a safety bulb is provided at the top of sealed capillary tube in order to protect thermometer.
- A calibrated scale is engraved on the surface of glass tube for indicating measured value of temperature.

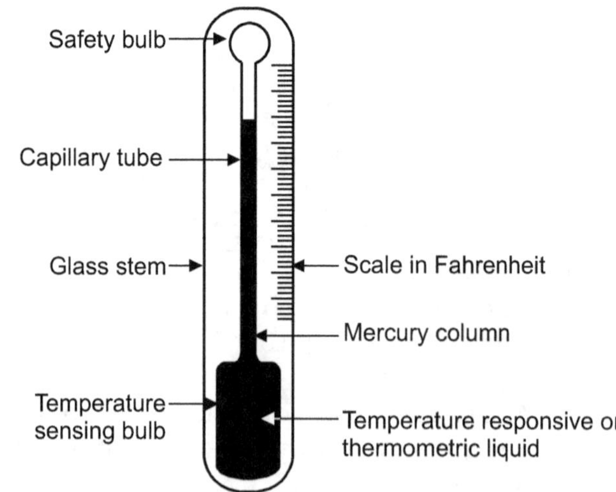

Fig. 2.15: Liquid in Glass Thermometer

- Size of capillary tube depends upon the size of sensing bulb, responsive liquid and desired temperature range of thermometer.

Working:

- When thermometer is placed near the hot body, whose temperature is to be measured, then responsive liquid contained in capillary tube will expand in volume.
- It causes the liquid to rise in capillary tube.
- As area of capillary tube is very small, increase in liquid volume will result in significant rise in liquid level.
- Length of movement of free surface of liquid column (Mercury column) will indicate the temperature, for which, calibrated scale is engraved on the surface of glass tube.

2.21.1 Precautions Taken for Accurate Readings with Minimum Dynamic Error

- Precautions to be taken for obtaining accurate readings with minimum dynamic error are,
    1. **Proper contact** between **sensing element and system**, whose temperature is to be measured.
    2. **No heat gain or heat loss** from the sensing element to the surrounding.
    3. **Sensing bulb** should have **good exposure** to **radiant energy** emitted by gas or liquid, whose temperature is to be measured.
    4. **Minimum thermal resistance** at sensing bulb.
    5. **Sensing bulb** should be **cleaned**, before use.
    6. **Accumulation** (deposition) of any **insulating film** over the **surface of sensing bulb** should be avoided.
    7. It should be ensured that, there is **no flow of air or gas** around the sensing bulb.

2.21.2 Desirable Characteristics or Properties of "Liquid in Glass Thermometer"

1. **High thermal conductivity of sensing element:**
    - If thermal conductivity of sensing element is low, then it will give slow response to temperature change. Therefore, thermal conductivity of sensing element should be high for fast response to temperature change.

2. **High heat transfer coefficient:**
    - Heat flow rate increases, if heat transfer coefficient of sensing element is high. It makes thermometer to give quick response.

3. **More surface area:**
    - Heat flow rate increases with increase in surface area per unit mass of sensing element. Like heat transfer coefficient, larger surface area also makes thermometer to give quick response.

4. **High velocity of atmospheric air surrounding the sensing element:**
    - Velocity of atmospheric air surrounding the sensing element should be high to get accurate reading of temperature under measurement.
    - If velocity of air passing over the sensing element is low, then there are chances of formation of a film around the sensing element.
    - Due to this film, heat flow rate decreases leading to slow rate of response. As air is passing slowly over the bulb, process of film formation will get enough time to cover sensing bulb and hence, thermometer will not show true value of temperature under measurement.

2.21.3 Temperature Range of Thermometric/ Responsive Liquids in a Glass Thermometer

Sr. No.	Liquid	Range (°C)
1.	Mercury	− 35 to 510
2.	Alcohol	− 80 to 70
3.	Toluene	− 80 to 100

2.21.4 Advantages of using Mercury as Thermometric/ Responsive Liquid in a Glass Thermometer

1. Large span between freezing temperature (− 38.53°C) and boiling temperature (356.73°C).
2. Linear coefficient of thermal expansion.
3. Easily available in a very pure state.
4. It does not stick to glass surface.
5. It is clearly visible through transparent glass.

2.22 ADVANTAGES, DISADVANTAGES AND APPLICATIONS OF "LIQUID IN GLASS THERMOMETER"

2.22.1 Advantages of "Liquid in Glass Thermometer"

1. Simple and easy to use.
2. Low cost.
3. Portability.
4. Good accuracy.
5. Physical damage (If any) of thermometer can be easily checked.
6. It is a self-operated system, because no auxiliary power is required for its operation.
7. No need of additional indicating instrument.

2.22.2 Disadvantages of "Liquid in Glass Thermometer"

1. Cannot be adopted for automatic recording.
2. Slow response with respect to change in temperature.
3. Time lag in measurement, i.e. time delay in sensing and indicating the temperature measured.
4. Range is limited to 300°C.
5. Not suitable for temperature measurement of any distant application. In other words, it cannot be adopted for remote reading.

2.22.3 Applications of "Liquid in Glass Thermometer"

1. Temperature measurement of liquids contained in open tanks, container etc.
2. Temperature measurement of cooking kettles, molten metal baths, pipe lines used for fluid flow, air ducts (pipes) etc.
3. To record temperature of standard room, calibration room etc.
4. Measurement of atmospheric temperature in industries, where, it is necessary to maintain a specific temperature during the occurrence of some processes.
5. Temperature measurement of human body.
6. If liquid inside sensing bulb is mercury, then "Liquid in glass thermometer" can measure temperature of any fluid having temperature up to 340°C. Mercury has boiling point temperature of 356.73°C. Hence, there is no possibility of evaporation of liquid mercury.

2.23 LIQUID IN METAL THERMOMETER (PRESSURE GAUGE THERMOMETER)

- **Method used:** Expansion of liquid due to increase in temperature.

Working principle:

- *If temperature increases, liquid expands in volume. This increase in volume of liquid is directly proportional to increase in temperature. Therefore, by measuring increase in volume of liquid, we can carry out temperature measurements.*

Construction:

- "Liquid in Metal Thermometer" or "Pressure Gauge Thermometer" consists of,

  1. **Pressure bulb:**
     - Pressure bulb is a **sensing element** containing **thermometer fluid (transmitting fluid or responsive fluid or thermometric fluid).**
     - **Pressure bulb** is a **metal tube having cylindrical shape.** One end of pressure bulb is closed and other end is connected to capillary tube. It is placed, where temperature is to be measured.

  2. **Capillary tube:**
     - It is a tube of very small diameter.
     - **Capillary tube** is made up of **copper or steel** for systems using fluids other than mercury.
     - When **mercury** is used as transmitting fluid, capillary tube is made up of **stainless steel.**
     - **Capillary tube** connects the **pressure bulb** to **transmitting system** and **readout device.**

  3. **Receiving element:**
     - It is a device, which responds to change in pressure or change in volume of thermometer fluid caused due to change in temperature.
     - Bourdon's pressure gauge or a bellow or a diaphragm can be used as a receiving element.
     - Receiving element is used to **convert change in pressure** or **change in volume** of **thermometer fluid (transmitting fluid)** into **displacement**.
     - Displacement so obtained may be amplified to operate a pointer over a stationary scale for indication of temperature.

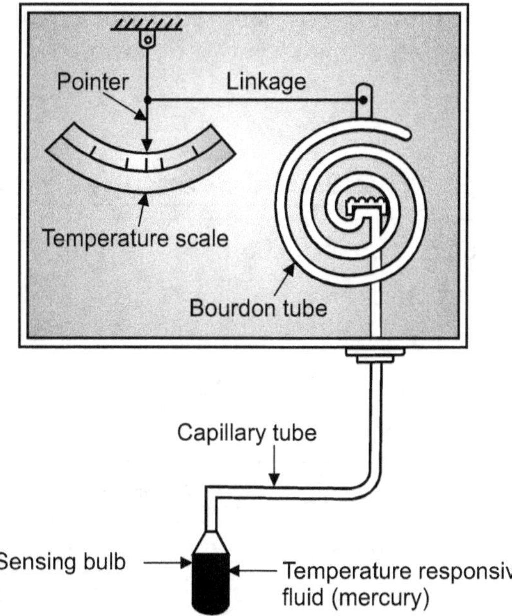

Fig. 2.16: Liquid in Metal Thermometer

2.23.1 Advantages of Pressure Gauge Thermometer

1. Rigid in construction.
2. Low cost.
3. Accuracy is good ($\pm$ 1% of full scale)
4. Fairly good response and sensitivity.
5. It is a self-operated system, because no auxiliary power is required for its operation.
6. If it is combined with a pneumatic or electrical transmission system for the purpose of magnification, then only auxiliary power is needed for its operation.
7. Less maintenance.

2.24 VAPOUR PRESSURE THERMOMETER

- **Operating Range:** with **SO₂** SO_2 (30°C to 120°C).

Working principle:

- *"**Vapour pressure** of a liquid depends upon the **temperature**. If this liquid is **partially filled** in an **enclosed chamber**, it will create a **definite vapour pressure**, by which, temperature can be measured".*
- Working of **Vapour pressure thermometer** is **similar** to **liquid filled thermometer**.

Construction & Working:

- Vapour pressure thermometer consists of a partially filled bulb.
- This bulb is used as sensing element to measure temperature.
- When bulb senses the temperature, vapour will be formed above the liquid level inside bulb.
- This vapour will pass through a capillary tube, and then to a bourdon tube (receiving element).
- Pressure of vapour formed causes the free end of bourdon tube to deflect.
- This deflection is transmitted by mechanical arrangement, causing the pointer to move over calibrated scale to give reading of measured temperature.

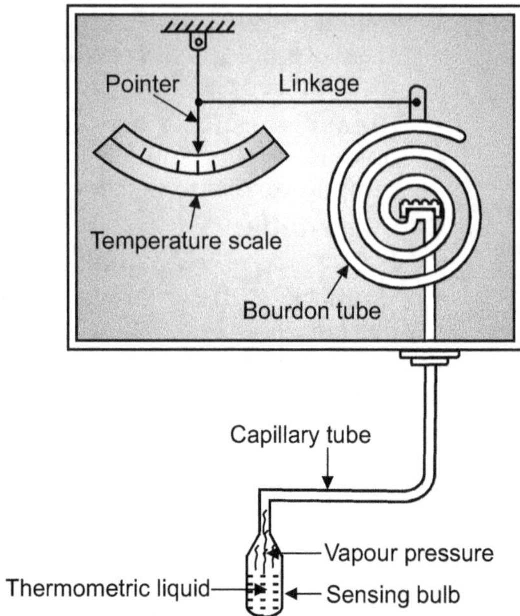

Fig. 2.17: Vapour Pressure Thermometer

2.24.1 Advantages of Vapour Pressure Thermometer

1. Rigid in construction.
2. Versatility.
3. Economical.
4. Easy to read from a remote distance.
5. Good response and sensitivity.

2.24.2 Disadvantage of Vapour Pressure Thermometer

1. Use of vapour pressure thermometer is limited due to availability of number of liquids, which can provide suitable vapour pressure ranges.

2.24.3 Applications of Vapour Pressure Thermometer

1. Industrial temperature measurements.
2. Commercial applications.

2.24.4 Reasons of Static Errors in Pressure Thermometers

1. Improper location of bulb.
2. Incorrect immersion of bulb.
3. Change in ambient pressure.
4. Change in ambient temperature.

2.25 DIFFERENCE BETWEEN NON-ELECTRICAL & ELECTRICAL METHODS

Sr. No.	Comparative Point	Non-electrical Methods	Electrical Methods
1.	Output signal	In this method, the output signal is **non-electrical** in nature.	In this method, the output signal is **electrical** in nature.
2.	Working principle	In this method, **expansion of liquid or metal** takes place with change in temperature.	In this method, **resistance or emf of material changes** with change in temperature.
3.	Precision & Accuracy	It is a **precise** method of temperature measurement.	It is an **accurate** method of temperature measurement.
4.	Cost	It is **economical** method of temperature measurement.	It is **costly** method of temperature measurement.
5.	Suitability	Best suited for **low temperature range.**	Best suited for **high temperature range.**
6.	Construction	The instruments have **simple** construction.	The instruments have **complex/complicated** construction.
7.	Examples or Types	Liquid in glass thermometer, Pressure thermometer, Bimetallic thermometer, Vapour pressure thermometer etc.	Thermocouple, Resistance thermometer (RTD), Thermistor, Radiation pyrometer etc.
8.	Control purpose	These methods are **not generally preferred** for temperature measurement to high accuracy.	These methods are **generally preferred** for temperature measurement, as they produce an electrical signal, which can be easily detected, amplified and used for **control purposes**.

2.26 ELECTRICAL METHODS OF TEMPERATURE MEASUREMENT

- Temperature measuring devices working on electrical methods of temperature measurement are called as **Electrical transducers**.
- These Electrical transducers convert the measured temperature into an equivalent electrical signal, such as, resistance in RTD and thermistor, whereas, voltage in thermocouple.
- These methods are very convenient, as they furnish an electrical signal. Electrical signal has the advantage that, it can be easily detected, amplified or used for control purposes.
- If temperature measuring devices working on electrical methods are properly calibrated, they give rise to highly accurate measurements.
- Methods used are of two types.
  (A) Using conductors (metals), and **(B) Using semiconductors.**
- Almost **all metals** used are of **PTC** type, whereas, **semiconductors** may be of **PTC** or **NTC** type.
- There are three types of temperature measuring instruments based on electrical concept.
  (a) Resistance Thermometer (RTD), (b) Thermistor, (c) Thermocouple

2.27 RESISTANCE THERMOMETER

Working Principle:
- Resistance thermometer works on the principle of positive temperature coefficient. This principle states that, **"electric resistance of metals is directly proportional to temperature, i.e. electric resistance of metals increases with increase in temperature."**
- Therefore, temperature of wire of any known material can be measured, if we measure its electrical resistance.

Construction:

- Sensing element is made up of metal having high positive temperature coefficient.

- Change in resistance caused by change in temperature is detected by the Wheatstone bridge.

- Materials are so selected that, values of Resistances R_1, R_2, R_3 and R_4 in four arms of bridge remain constant under a normal temperature. At this condition, galvanometer shows zero reading. Refer Fig. 2.18 (a).

- Galvanometer has a pointer placed in the plane of circular scale, which is directly calibrated in terms of temperature.

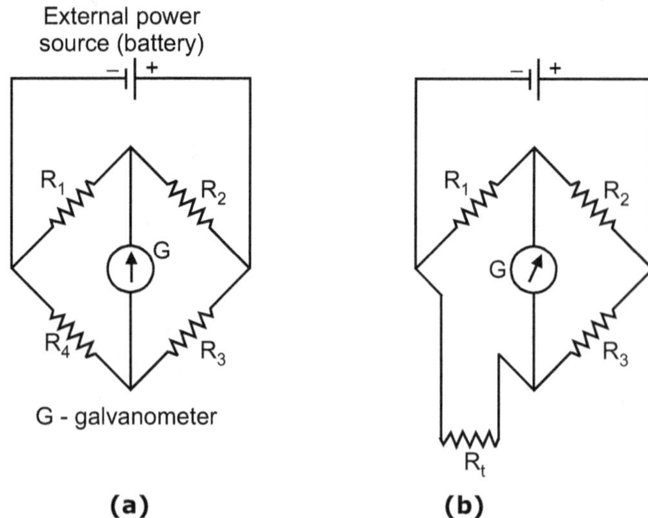

Fig. 2.18: Use of Wheatstone Bridge

- When no current flows through the galvanometer, then according to normal principle of Wheatstone bridge,

$$\frac{R_1}{R_3} = \frac{R_4}{R_2}$$

- Now, fourth arm having resistance R_4 is replaced by RTD's sensing element, i.e. sensing wire.

- Let R_t be the resistance of sensing wire. Due to change in temperature, resistance R_t changes and Wheatstone bridge becomes unbalanced.

$$\frac{R_1}{R_3} = \frac{R_t}{R_2}$$

- This will make the pointer to move over the circular scale of galvanometer, which is calibrated in terms of temperature. Position of pointer indicates the value of temperature measured. Refer Fig. 2.18 (b).

2.28 PLATINUM RESISTANCE THERMOMETER (RTD)

Operating Range: – 200°C to 1000°C.

Working Principle:

- *Positive temperature coefficient of resistance (PTC).*

Construction:

- Modern platinum resistance thermometer consists of pure platinum wire wound on hollow pipe made up of insulating mica or ceramic, which is placed in porcelain sheath. Refer Fig. 2.19.

- Free ends of platinum wire are attached to long leads of low resistance copper wires.

- To measure change in resistance, Wheatstone bridge is used.

- Two long extension leads of RTD form one arm of Wheatstone bridge circuit. Refer Fig. 2.18.

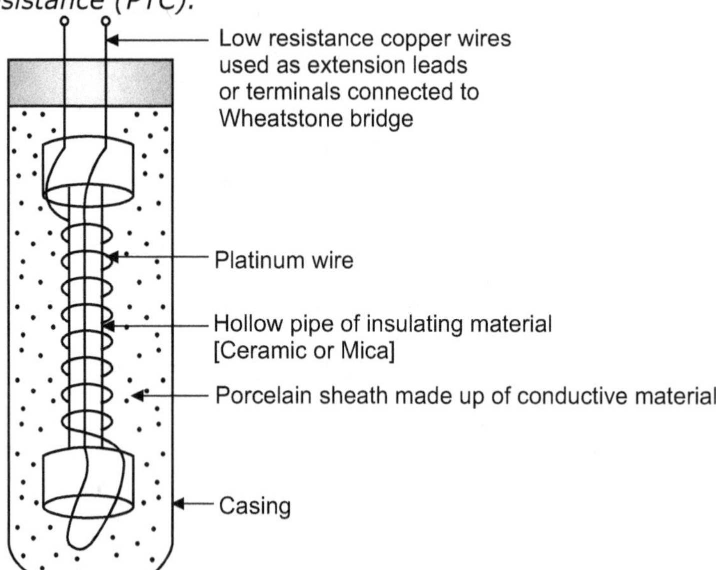

Fig. 2.19: Platinum Resistance Thermometer (RTD)

Working:

- When RTD is subjected to temperature variation, the Wheatstone bridge gets unbalanced due to change in resistance of RTD arm.
- This makes the pointer to move over circular scale of galvanometer, which is directly calibrated to give measured value of temperature.

2.29 RTD MATERIALS

2.29.1 Essential Characteristics for RTD Material

1. Change in resistance of material should be large enough for unit change in temperature.
2. High resistivity, so that, minimum volume of material can be used for construction of RTD.
3. Resistance of material should have a continuous and stable relationship with temperature.

2.29.2 Materials Used for RTDs

Sr. No.	Metal	Operating Temperature Range (°C)	
		Minimum	Maximum
1.	Platinum	− 260	110
2.	Copper	70	120
3.	Nickel	− 20	200
4.	Tungsten	− 200	1000

2.29.3 Advantages of using Platinum in RTD

1. Linear relationship of Resistance with temperature.
2. Platinum gives stable value of resistance within its normal operating temperature range. Therefore, output device shows reliable reading of temperature measured.
3. Chemically inert.
4. It gives sufficient change in resistance, which can be easily measured and calibrated to give temperature over wide range.

2.30 ADVANTAGES, DISADVANTAGES AND APPLICATIONS OF RTDS

2.30.1 Advantages of RTDs

1. Wide operating range from − 200°C to 1000°C.
2. Easy to install and replace.
3. Best suited for remote indication.
4. Better sensitivity.
5. Extremely accurate in sensing, measuring and displaying the temperature measured.
6. Stability of performance over a long period of time.
7. Linear Resistance-Temperature relationship.

2.30.2 Disadvantages of RTDs

1. High cost.
2. Needs bridge circuit with power supply.
3. Possibility of self-heating.
4. Large bulb size required for measuring high temperatures.
5. Sometimes, balancing of bridge takes more time.

2.30.3 Applications of RTDs

1. Temperature measurement in heating ovens, drying ovens, pressure vessels, baths etc.
2. Temperature measurement of cold storage plant, boiler and power generator etc.
3. For measurement of radiant heat.
4. Food processing plants.
5. Temperature measurement of exhaust gases.
6. Petrochemical industries.
7. Refrigeration and Air-conditioning applications.
8. Process industries.

Note:

1. If R_0 = Resistance of RTD measured at temperature T_0 (i.e. Temperature 0°C)

   R_T = Resistance of RTD measured at temperature T, when it comes in contact with hot surface or medium, and

   α_0 = Temperature coefficient of material, then,

   Resistance "R_T "measured at temperature "T" in °C can be calculated by,

   $$R_T = R_0 [1 + \alpha_0 (T - T_0)]$$

2. Unknown temperature "T" is calculated by,

   $$T = \left(\frac{R_T - R_0}{R_{100} - R_0}\right) \times 100$$

   where, R_{100} = Resistance of wire at temperature 100°C

   R_0 = Resistance of wire at temperature 0°C

   R_t = Resistance of wire at temperature T°C

2.31 THERMISTOR

Operating Range: – 150°C to 300°C.

- The term "thermistor" means **Thermal – Resistor**.
- It is also called as, **"Thermally Sensitive Variable Resistor".**

Working Principle:

- Thermistors work on following principles.

  (a) **Negative temperature coefficient (NTC),** where resistance of material decreases with increase in temperature. **Materials** possessing the property "NTC" are composed of **metallic oxides (semiconductors)** such as, **Manganese, Nickel, Cobalt, Copper, Iron and Uranium.**

  (b) **Positive temperature coefficient (PTC)**, where resistance of material increases with increase in temperature. **Materials** possessing the property "PTC" are composed of **metals and metal oxides,** such as, **Powdered Barium Carbonate, Barium Titanate and Titanium Oxide.**

- In both cases, Relationship of **Temperature** Versus **Resistance** is **non-linear**, i.e. exponential.

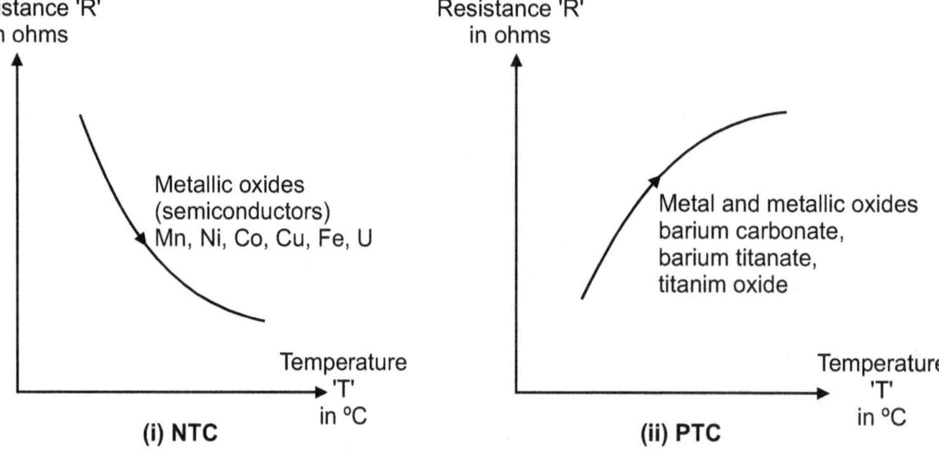

Fig. 2.20: Working Principle of Thermistors

Construction:

- **Thermistors** working on the **principle of NTC** are preferred in **industrial applications.**
- Thermistor working on the principle of NTC is also called as **"Temperature Sensitive Semiconductor".**
- Thermistors of NTC type are made up of semiconductor materials. They are manufactured by sintering the mixtures of metallic oxides, such as Manganese, Nickel, Cobalt, Copper and Uranium.
- These metallic oxides are mixed in suitable proportions with binders and compressed into desired shapes and sizes. Therefore, thermistors are available in various sizes and shapes to suit any particular application.
- Thermistors have lead for connectivity with circuits.
- Fig. 2.21 shows **various forms or shapes of thermistors**, such as, bead, probe, disk, rod, washer type etc.

   (i)   Bead type thermistor:
- o   Bead type thermistor is smallest in size having diameter in the range of 0.01 mm to 1.2 mm.
- o   Bead type thermistor is sealed in the tips of solid glass rods to form probe type thermistor.

   (ii) Probe type thermistor:
- o   Probe type thermistors have diameter 2.5 mm (approx.) and length in the range of 6 mm to 50 mm.

   (iii) Disc type thermistor:
- o   Material used for manufacturing disc type thermistor is pressed under high pressure to obtain it in the form of cylindrical shape having diameters from 2.5 mm to 25 mm.
- o   Therefore, a large variety of disc type thermistors having different diameters is available in market. It makes possible to select type of thermistor to suit any particular type of application.

   (iv) Washer type thermistor:
- o   Design of disc type thermistors is modified to obtain washer type thermistors (washer shaped).

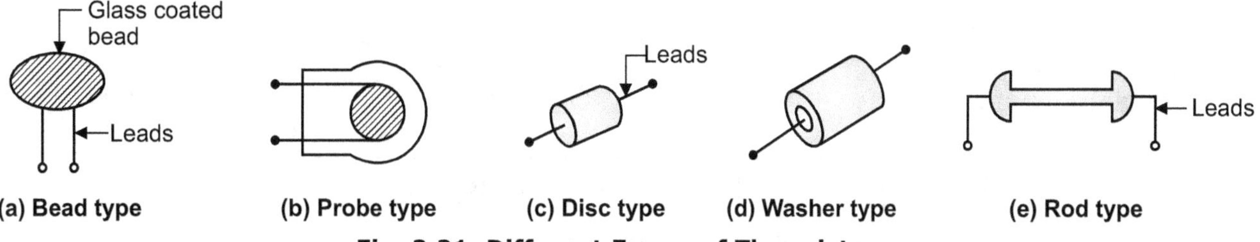

Fig. 2.21: Different Forms of Thermistors

2.32 ADVANTAGES, DISADVANTAGES AND APPLICATIONS OF THERMISTOR

2.32.1 Advantages of Thermistor

1. It is more stable.
2. Small in size.
3. Low cost.
4. High sensitivity.
5. Fast thermal response.
6. Stronger and rigid construction.
7. Easily adaptable to electrical read out devices.

2.32.2 Disadvantages of Thermistor

1. It requires external energy source.
2. Not suitable for wide range of temperature.
3. Need of shielded power lines.
4. Cannot be used to measure temperature higher than 300°C.
5. Non-linear resistance-temperature relationship.

2.32.3 Applications of Thermistor

1. Used as sensor in measurement systems of liquid level, pressure, flow etc.
2. Used for measurement of power at high frequency.
3. Used for temperature control, temperature compensation and time delay circuits.
4. Used for measurement of temperature in Vacuum.
5. Used for measurement of temperature of composition of gases.
6. Used for measurement of thermal conductivity.

2.33 INDUSTRIAL OR PRACTICAL APPLICATIONS OF THERMISTORS

2.33.1 Use of Thermistor for Temperature Control

- A sensitive temperature control can be achieved by placing thermistor in one arm of a Wheatstone bridge, a variable resistor in another arm and a polarized relay across the output. Refer Fig. 2.22.
- When temperature around the thermistor increases, it becomes warm and operate the relay in one direction.
- Conversely, when temperature around the thermistor decreases, it becomes cool and operate the relay in opposite direction.
- Point of temperature control and operation can be changed by changing the value of resistance in variable resistor.

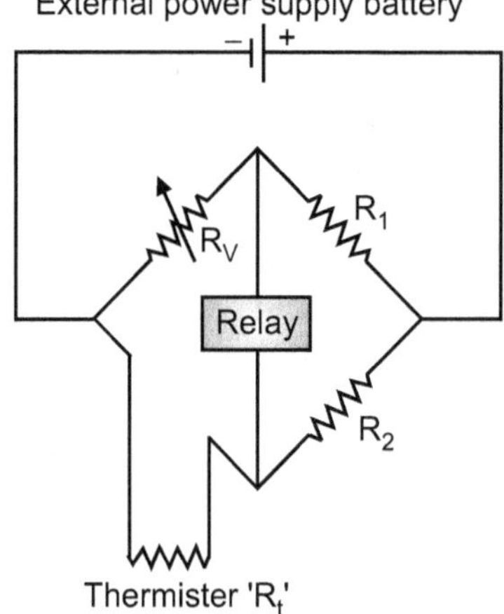

Fig. 2.22: Use of Thermistor for Temperature Control

2.33.2 Use of Thermistor for Liquid Level Control

- Thermistor can be used as liquid level controller. For this purpose, thermistor is immersed in a liquid tank.
- Thermistor is connected with a relay and battery as shown in Fig. 2.23.

Case (i): Liquid level decreases below desired level

- When level of liquid decreases below the location of thermistor, then thermistor will be exposed to atmosphere. As thermistor comes in direct contact with atmospheric air, its temperature will increase.
- Due to increase in temperature, resistance of thermistor will decrease. This will permit enough current to flow through the circuit, and energises the relay to start the liquid pump. Thus, liquid comes in tank and its level rises up to desired level.

Case (ii): Liquid level increases causing overflow

- When level of liquid increases, thermistor will be submerged in liquid. Thermistor will be cooled and its temperature will decrease.

- Due to decrease in temperature, resistance of thermistor will increase. This will reduce the flow of current through the circuit, which makes the relay to stop the liquid pump. This will avoid any possibility of overflow of liquid.
- Thus, thermistor controls and maintains the desired liquid level in tank.

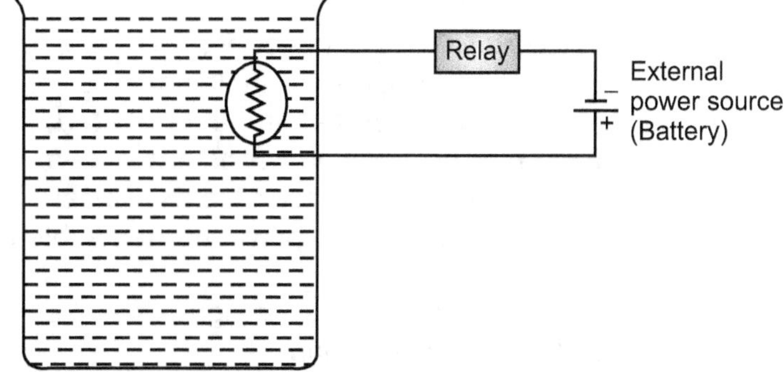

Fig. 2.23: Use of Thermistor as Liquid Level Controller

2.34 CHARACTERISTICS OF THERMISTOR

1. **Resistance - Temperature characteristic:**
   - Thermistor follows the relation,

$$R_t = a. \ e^{\left(\frac{b}{T}\right)}$$

Where, "a" & "b" are constants,

R_t = Resistance of thermistor at absolute temperature "T".

2. **Voltage - Current characteristic:**
   - If voltage is gradually increased, the current will increase and the heat generated in thermistor will finally begin to raise its temperature above surrounding temperature.

3. **Current - Time characteristic:**
   - Current - Time effect provides a simple and accurate way of achieving time delays from milli-seconds to many minutes.

2.35 WORKING PRINCIPLE OF THERMOCOUPLE

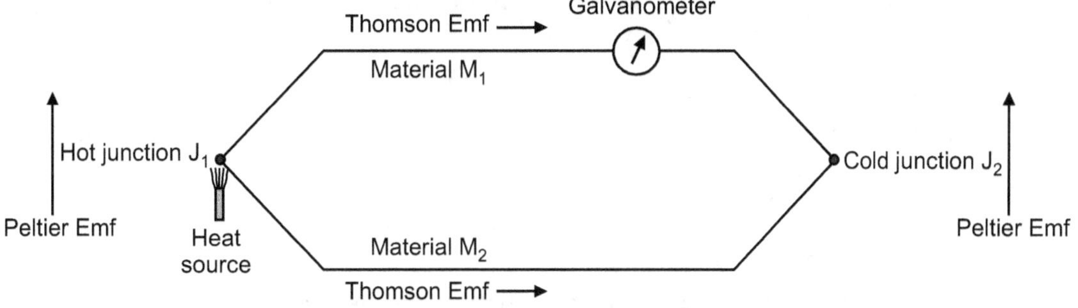

Fig. 2.24: Working Principle of Thermocouple

Seebeck Effect:
- It states that, ***"when two dissimilar metal wires are joined to form a complete electric circuit and the two junctions are maintained at different temperatures, an electromotive force (emf) is set up to establish flow of current."***
- Magnitude of emf generated is directly proportional to magnitude of temperature difference of two junctions and materials used for wires or conductors.
- Let us consider two wires made up of different materials. Ends of both wires are joined to form two contact points or junctions.

Net Emf Generated:
- Following emfs will be generated due to Thomson effect and Peltier effect.
  1. **Thomson effect:**
     - According to Thomson effect, if one end of wire is heated, then a Thomson emf is generated or developed between two ends of that single wire.
     - **Two Thomson emfs** will be generated, since there are two wires.

2. Peltier effect:
   o According to Peltier effect, when a junction formed due to contact of ends of two wires made-up of dissimilar metals is heated or cooled, a Peltier emf is generated.
   o **Two Peltier emfs** will be generated, since there are two junctions.
- Therefore, **Net emf** will be **resultant** of **two Peltier emfs** and **two Thomson emfs**.
- However, in commercial instruments, materials for wires or conductors are so selected that, emfs due to Thomson effect will be negligible and the instruments will measure Peltier emfs only.
- It is important to note that, magnitude of emf generated by Peltier effect is proportional to temperature of hot junction.
- *If temperature at one junction (reference junction, i.e. cold junction) is kept constant, then effective emf will be generated due to change in temperature at another junction, i.e. hot junction. Temperature difference between hot and cold junction can be measured by this effective emf generated in thermocouple.*
- Therefore, **hot junction** is called as **measuring end**, where **temperature is to be measured** and **cold junction** is called as **reference junction**, where **reading of temperature difference** between two junctions **is taken.**

2.36 THERMO-ELECTRIC LAWS

1. "Application of heat to single homogeneous metal is not capable of producing or sustaining electric current."

2. **Law of Intermediate Metals:** "Introduction of third metal into a thermocouple circuit will have no effect on the emf generated, as long as, the junctions of third metal with the thermocouple metals are at same temperature."

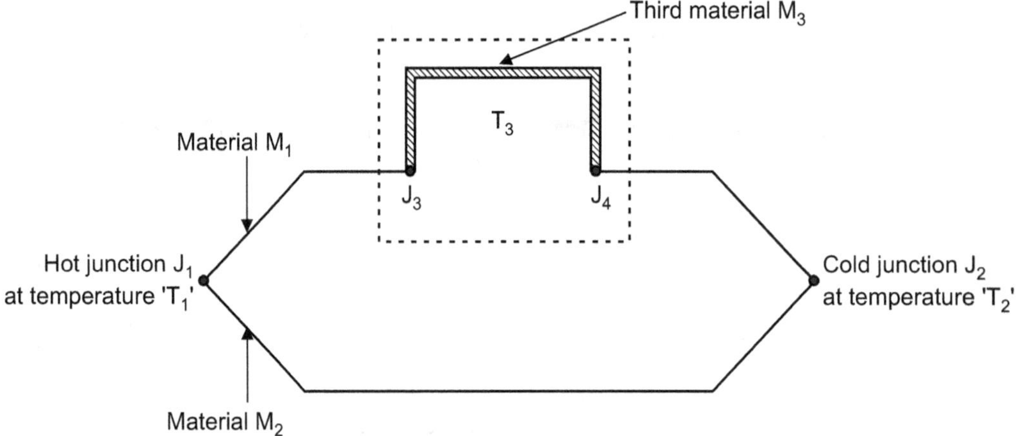

Fig. 2.25: Law of Intermediate Metals

   o Fig. 2.25 shows a circuit consisting of two dissimilar homogeneous metals having junctions 'J$_1$' and 'J$_2$', which are at higher temperature 'T$_1$' and lower temperature 'T$_2$' respectively.
   o Due to temperature difference between two junctions, an emf will be developed. This value of emf is a measure of that temperature difference.
   o This value of emf developed will not be affected, if a third homogeneous metal is introduced as a part of circuit, provided that, temperatures of both new junctions (J$_3$ and J$_4$ formed) are at same temperature 'T$_3$'.

Applications of Law of Intermediate Metals:

(i) This law makes it possible to use extension wire of a metal different from thermocouple material. For example: Inexpensive and cheaper material, such as, Copper can be used as **Extension wires/leads**.

(ii) This law enables **introduction (addition)** of a **measuring instrument (Galvano-meter)** into the circuit to measure the emf produced. Refer Fig. 2.24.

3. Law of Intermediate Temperature: "If a thermocouple circuit develops an emf e_1, when its junctions are at temperature T_1 and T_2 and an emf e_2, when its junctions are at temperature T_2 and T_3, then it will develop an emf equal to $(e_1 + e_2)$, when its junctions are temperature T_1 and T_3." Refer Fig. 2.26.

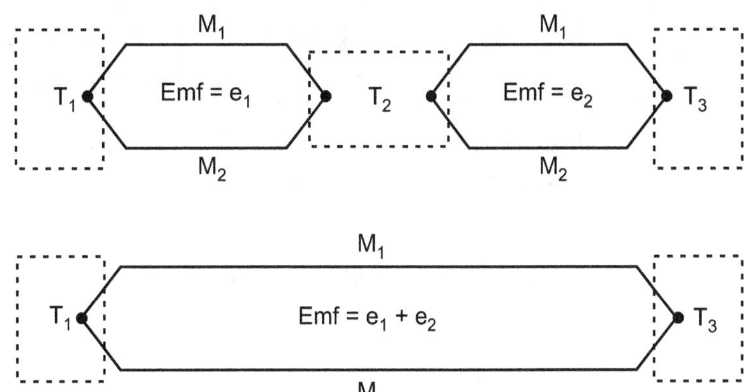

Fig. 2.26: Law of Intermediate Temperature

Applications of Law of Intermediate Temperature:

- This law is used to make correction to thermocouple reading, if the reference junction temperature is different from the temperature, at which, thermocouple was calibrated.
- For this purpose, manufacturer companies producing thermocouples provide calibration curves, charts and tables for reference junction temperature of 0°C along with thermocouple.
- Now, if a thermocouple (calibrated with a reference junction temperature at 0°C) is used with reference junction at 20°C, then the emf reading obtained is to be added by lost emf due to reference junction being at 20°C.
- This value of lost emf can be taken from the table of correction provided along with thermocouple.

2.37 THERMOCOUPLE (THERMO-ELECTRIC PYROMETER)

Operating Range: 200°C to 1600°C.

Construction:

- Thermocouple consists of two dissimilar metal conductors, which are electrically insulated except the hot junction. Refer Fig. 2.27.
- At hot junction, the conductors may be soldered or welded together.
- A refractory/metal sheath is used to protect thermocouple from mechanical damage.
- Two metal wires are extended as compensating leads or extension leads, which allows the meter or read-out device (Reference junction) to be placed at a considerable distance away from hot junction of thermocouple.

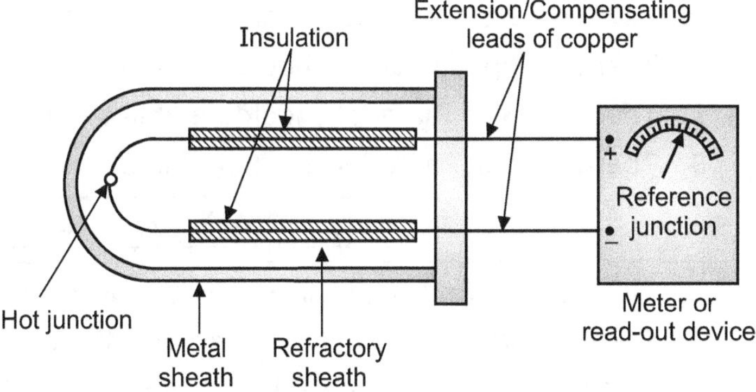

Fig. 2.27: Thermocouple

Working:

- Thermocouple is mostly used to measure high temperatures in the furnaces.
- Junction kept near the hot source (furnace) is called as **hot junction.**
- Junction, at which, emf generated is measured, is called as **cold junction (reference junction).**
- The magnitude of emf generated depends upon the temperature difference of two junctions and type of metals used.
- Emf generated is directly calibrated in terms of temperature at read-out device or meter reader to indicate reading of temperature difference between two junctions.

2.37.1 Factors to be Considered while Selecting Thermocouples for Temperature Measurement

- Various considerations in selection of thermocouples for measurement of temperature are:
  1. Construction, whether rigid.
  2. Strength.
  3. Atmospheric or ambient conditions.
  4. Type of insulation used.
  5. Sensitivity.

2.37.2 Desirable Characteristics of Thermocouple

1. Emf generated per degree change in temperature must be sufficient for detection and measurement purpose.
2. Temperature-EMF relationship should be linear and reproducible.
3. Thermocouple should maintain its calibration without drift over a long period of time.
4. Thermocouple should have a long life, so that, cost of measurement will be less for less frequent measurements.
5. Thermocouple material should be able to withstand high and rapidly fluctuating temperatures.
6. Thermocouple material should be highly resistant to oxidation, corrosion and contamination.

2.37.3 Desirable Properties of Thermocouple Material

1. Thermocouple material should be able to give fast thermal response.
2. Thermocouple material must have stability over a wide temperature range.
3. Thermocouple material should not produce drift.
4. Thermocouple material should have long life.
5. Temperature-EMF relationship should be linear and reproducible.
6. Thermocouple material should be able to withstand against high temperatures.
7. Thermocouple material should be able to withstand against rapidly fluctuating temperatures.
8. Thermocouple material should be highly resistant to oxidation, corrosion and contamination.

2.37.4 Thermocouple Materials

- Commonly used materials are **Copper, Iron, Platinum, Rhodium and Alloys** like **Constantan** (60% Cu + 40% Ni) and **Chromyl** (10% Cr + 90% Ni).
- Size of wire depends upon the temperature range to be measured.
- For measurement of high temperature, heavier wires are used.

Sr. No.	Metal	Temperature Range (°C)	
		Minimum	Maximum
1.	**Iron/Constantan**	– 200	900
2.	**Chromyl/Alumel**	– 200	1250
3.	**Platinum/Rhodium**	0	1600
4.	**Copper/Constantan**	– 200	400

2.38 ADVANTAGES, DISADVANTAGES AND APPLICATIONS OF THERMOCOUPLE

2.38.1 Advantages of Thermocouple

1. Wide measurement range of temperature from – 200°C to 1600°C.
2. Cheaper than resistance thermometer (RTD).
3. Accurate measurements.
4. Good reproducibility.
5. Low cost.
6. No need of external supply.
7. Rigid and robust in construction, so it can be used in high vibration or adverse environments.
8. Bridge circuits are not needed for measurements.
9. Ability to get connected to suitable electrical equipment for measuring rapid or sudden changes in temperature.

2.38.2 Disadvantages of Thermocouple

1. Thermocouple does not allow any compromise in essential requirements during temperature measurement, such as, Cold junction, Cable and Lead compensation.
2. Output is very small, therefore, it needs signal amplification.
3. At high temperatures, thermocouple shows non-linear relationship of **Temperature-EMF**.
4. Not suitable in areas subjected to high radiation fields.
5. Slower in response as compared to RTD. Therefore, it cannot be used for high sensitivity applications.
6. Proper separation of extension leads from thermocouple wires is essential to avoid errors.

2.38.3 Applications of Thermocouple

1. To monitor temperatures of liquids and gases in flowing pipes, ducts or storage.
2. Industrial furnaces, hot baths, solutions in chemical reactors etc.
3. For temperature measurement in cryogenic range (−150°C to 273°C).
4. Measurement of subzero temperature in refrigeration system.

2.39 DIFFERENCE BETWEEN RTD AND THERMOCOUPLE

Sr. No.	Comparative Point	RTD or Resistance Thermometer	Thermocouple
1.	**Working principle**	RTD directly measures the actual temperature of object, because electrical resistance is directly proportional to temperature.	Thermocouple measures the temperature difference between hot junction and cold junction.
2.	**Material used**	Metals like Copper, Platinum, Nickel etc.	Metals like Copper, Iron, Platinum, Rhodium or Alloys Like Constantan.
3.	**Output reading**	Actual temperature of object.	Temperature difference between temperature of cold junction and hot junction.
4.	**Temperature range**	Low range of measurement from – 200°C to 1000°C.	Wide range of measurement from – 200°C to 1600°C.

contd. ...

5.	Sensitivity	RTD is more sensitive than thermocouple.	Less sensitive due to very small emf output.
6.	Effect of change in atmospheric temperature	No effect on accuracy of measurement.	Affects accuracy of measurement.
7.	Response	Better response.	Comparatively slow response.
8.	Cost	High cost.	Moderate cost.

2.40 DIFFERENCE BETWEEN THERMISTOR AND THERMOCOUPLE

Sr. No.	Comparative Point	Thermistor	Thermocouple
1.	Material used	Metallic oxides of Cu, Ni, Fe.	Metals like copper, iron, platinum, rhodium or alloys like constantan.
2.	Output reading	Actual temperature of object.	Temperature difference between temperature of cold junction and hot junction.
3.	Operating or Temperature range	Narrow operating range from $-150°C$ to $300°C$.	Wide operating range from $-200°C$ to $1600°C$.
4.	Construction	Complex.	Simple, but rigid.
5.	Size	Very small.	Comparatively much larger.
6.	Cost	Expensive.	Cheaper.
7.	Effect of environmental conditions	No effect on accuracy.	Accuracy gets affected.
8.	Sensitivity	Extremely sensitive. It rapidly measures temperature of object, because, temperature change is directly proportional to change in resistance of sensing element.	Considered to be less sensitive, because of very small emf output.
9.	Response	Better response.	Comparatively slow response.
10.	Need of insulation	No. Insulation covering is not required, during temperature measurement of any conducting fluid.	Yes. Insulation covering is required, during temperature measurement of any conducting fluid.
11.	Applications	(a) Used for temperature measurement with control. (b) Used for controlling liquid level.	Used for temperature measurement only.

2.41 DIFFERENCE BETWEEN RTD AND THERMISTOR

Sr. No.	Comparative Point	RTD or Resistance Thermometer	Thermistor
1.	Material used	Metals like Copper, Platinum, Nickel etc.	Metallic Oxides of Cu, Ni, Fe, Mn etc.

contd. ...

2.	**Working principle**	Positive Temperature Coefficient of Resistance.	Positive as well as Negative Temperature Coefficient of Resistance.
3.	**Operating temperature range**	– 200°C To 1000°C.	– 150°C To 300°C.
4.	**Accuracy**	More.	Comparatively less accurate.
5.	**Precision**	Less precise.	Highly precise.
6.	**Reproducibility**	Better than thermistor.	Comparatively less.
7.	**Hysteresis**	Low.	More.
8.	**Size**	Large.	Comparatively small.
9.	**Cost**	High.	Comparatively less.
10.	**Construction**	Complex in construction.	Easy to construct.
11.	**Life / Durability**	Less durable.	More durable.
12.	**Applications**	Used for temperature measurement only.	(i) Used for temperature measurement with control. (ii) Used for controlling liquid level.

2.42 QUARTZ THERMOMETER OR QUARTZ CRYSTAL THERMOMETER

Working principle:

- Quartz Crystal Thermometer works on the **principle of piezoelectric effect**, where a "Resonant frequency change" occurs due to "Temperature change" of the Crystal/Integrated circuit.

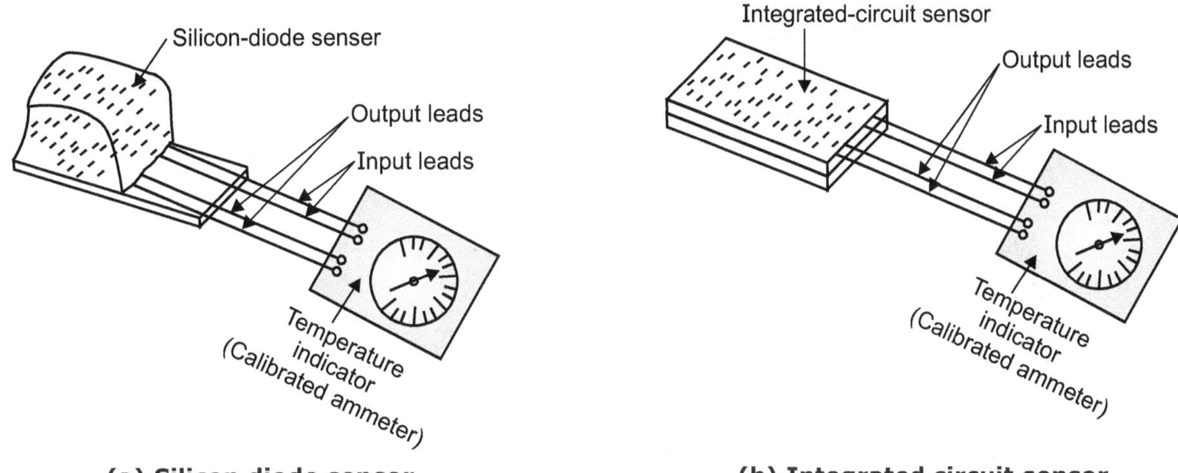

(a) Silicon diode sensor   **(b) Integrated circuit sensor**

Fig. 2.28: Schematic Arrangement of Quartz Thermometer

Construction and Working:

- A piezoelectric crystal provides a highly accurate and sensitive method of temperature measurement based on the change in its resonant frequency, which is directly proportional to the temperature change.
- The resonant frequency of a quartz crystal plate depends upon the temperature.
- When the crystal is cut at a proper angle or crystal is cut in the form of shear type LC cut, the change in frequency is linearly related to the change in the temperature of the crystal.
- In other words, when proper cut is given to quartz crystal, we get linear relationship of change in frequency with change in temperature of crystal.

- This linear relationship (i.e. linearity) is utilized to obtain highly accurate measurement readings of temperature.
- Here, a "nominal resonator frequency" of 28 MHz is supplied by a "reference - oscillator" to the "temperature sensor".
- Due to temperature imposed to the Quartz crystal (silicon diode sensor), its resonance frequency increases. This "frequency - difference" is detected by the associated electronic circuitry of this thermometer, which consists of frequency counters and digital read-outs.

- Thus, the detected "frequency - difference" is converted to pulses and then passed to an electronic counter, which provides a digital display of the temperature magnitude.
- The fundamental frequency "f_o" depends upon the thickness of the crystal and it can be adjusted so as to give a sensitivity of the order of 1000 Hz for a temperature change of 1°C.
- In other words, the detection of change in frequency of oscillation of 1 Hz gives a resolution of 0.001°C. Thus, quartz thermometer has better resolution.
- The temperature range of measurement is about − 40°C to + 230°C with an accuracy of ± 0.1% and excellent repeatability.

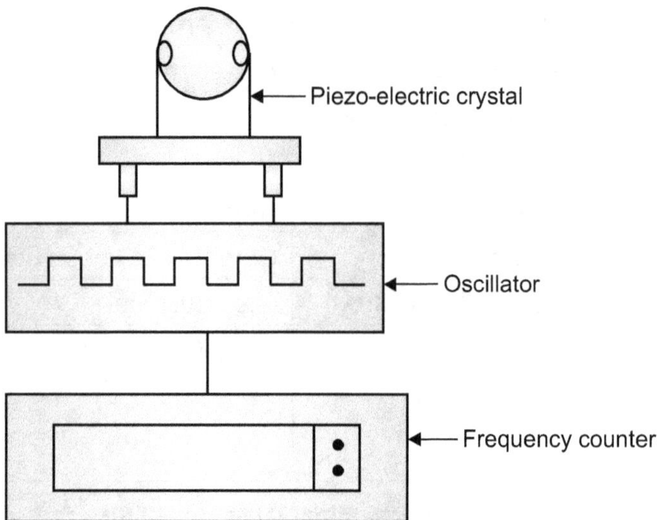

Fig. 2.29: Associated Electronic Circuitry of Quartz Thermometer

2.42.1 Advantages and Limitations of Quartz Thermometer

Advantages of Quartz Thermometer:
1. Highly linear output.
2. Long-term stability and reliability.
3. High resolution of order 0.001°C can be obtained.
4. Excellent repeatability.
5. High accuracy of the order of ± 0.1 %.
6. Excellent repeatability in the measuring range of − 40 to 230°C.
7. Fast response.
8. Remote sensing up to 3000 m is possible.

Limitations of Quartz Thermometer:
1. Limited measuring range, i.e. − 40°C to 230°C.
2. Piezoelectric crystals have strong cross-sensitivity for pressure changes, if they occur simultaneously in the temperature measuring systems.

2.43 INTRODUCTION TO PYROMETRY AND PYROMETER

- **Pyrometry** is **contact-less method** of temperature measurement. It is the technique of temperature measurement of a body by measuring its electromagnetic radiations.
- **Pyrometer** is used, when temperature being measured is very high and physical contact is impracticable or impossible.
- At temperatures above 650°C, radiations emitted by hot body have sufficient intensity, so that, technique of Pyrometry can be used for temperature measurement.
- Temperature measuring instruments, such as, thermocouple, thermistor, thermometer etc. may melt due to their physical contact with the hot body having very high temperature.

Working Principle:

- When a body is heated, it emits thermal energy in the form of electromagnetic radiations. This property is used as working principle of Pyrometer.
- Working of pyrometer depends upon the **relationship** between **Electro-magnetic radiation emitted by hot body** and **temperature of that hot body**.

2.44 RADIATION PYROMETER OR "TOTAL RADIATION PYROMETER"

Operating Range: More than 750°C.

Working Principle:

- Radiation pyrometer works on the principle, which states that, **"Temperature of a body can be measured by measuring radiant energy emitted by that hot body."**
- According to Stefan Boltzmann's law, total thermal radiation emitted by a body can be calculated by,

$$Q = \sigma \cdot T^4 \qquad \text{in Watt/m}^2 \text{ or W.m}^{-2}$$

Where, σ = Stefan Boltzmann's constant   in $W.m^{-2} K^{-4}$

And T = Absolute temperature   in Kelvin

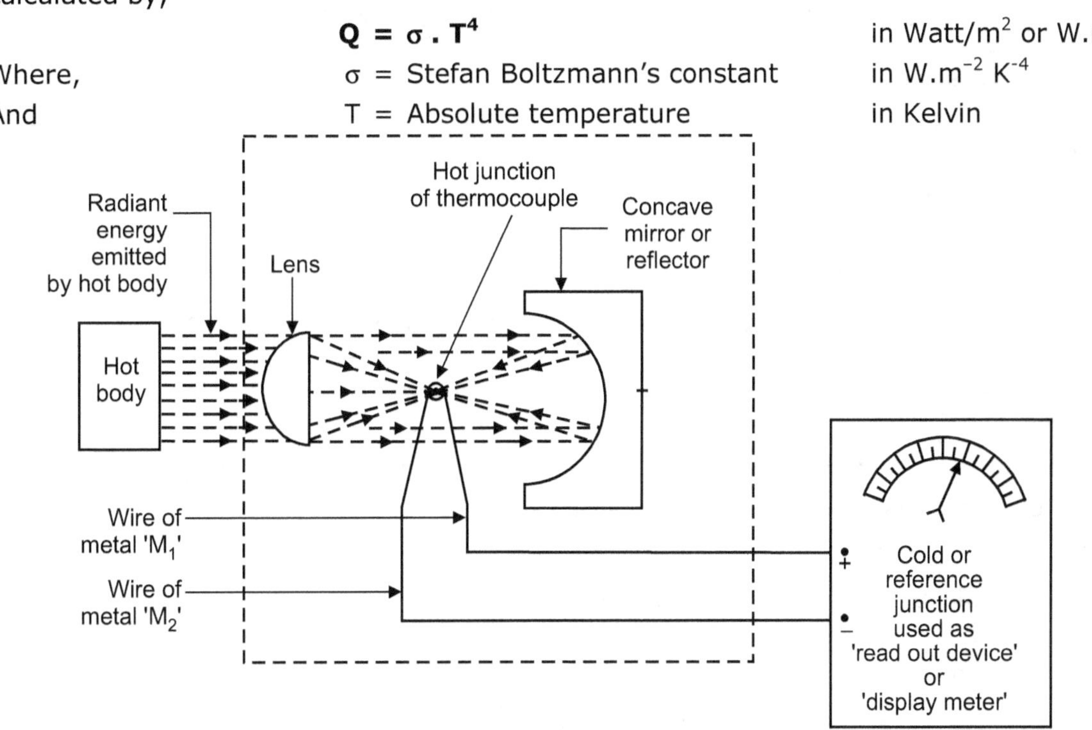

Fig. 2.30: Total Radiation Pyrometer

Construction:

- Radiation pyrometer consists of three major components:

    (a) Lens: It is used to focus the radiated energy from the body (whose temperature is to be determined) on to a receiving element.

    (b) Receiving element: Resistance thermometer or thermocouple can be used.

    (c) Recording instrument: A temperature indicator, recorder or controller is attached with the receiving end to obtain reading.

Working:

- When the total energy radiated by a hot body enters the pyrometer, it is collected and focused by lens on to the receiving element (thermocouple).
- Due to hot radiations incident on thermocouple, an emf is developed. This emf is measured by a milli-voltmeter, which is calibrated to show value of measured temperature directly.
- Concave mirrors are used to avoid reflection of radiations in unwanted directions.
- Concave mirrors reflect them back to receiving element as illustrated in Fig. 2.30.

2.45 ADVANTAGES, DISADVANTAGES AND APPLICATIONS OF RADIATION PYROMETER

2.45.1 Advantages of Radiation Pyrometer

1. Ability to measure high temperature, unlike thermocouple, which may melt at high temperatures.
2. No physical contact with object, whose temperature is to be measured.
3. Fast speed of response.
4. High output and moderate cost.
5. It is less affected by corrosive atmosphere.
6. High accuracy of ± 2 % of full scale.

2.45.2 Disadvantages of Radiation Pyrometer

1. Non-linear scale.
2. Possible errors due to presence of intervening gases or vapours, which absorb the radiations.
3. Emissivity of hot body affects temperature measurement.
4. Due to dirt or dust on the mirror or lens, instrument shows low value of temperature measured than actual temperature of body.

2.45.3 Applications of Radiation Pyrometer

1. It is preferred for temperature measurement of components or applications like furnace interiors, boilers etc., which are not easily accessible.
2. It measures temperatures of places or components, which are not easy to access.
3. It is used to measure average temperature of large surfaces.
4. It is used to measure temperature of moving object or target.
5. It is used to measure invisible rays from radiations.
6. It is used to measure temperature in corrosive atmospheric conditions, where other temperature measuring instruments may be contaminated.

2.46 OPTICAL PYROMETER

Operating Range:

- Up to 1400°C, which can be extended up to 3000°C with the use of absorption type screen placed near the lens.

Working Principle:

- It is based on comparisons of intensity (brightness) of visual radiation emitted by a hot body with the radiation energy emitted by a source of known intensity.
- Here, the intensity of radiation emitted by hot body, whose temperature is to be measured, is matched with the intensity of a reference filament (lamp), whose temperature is known.

Construction & Working:

- Optical pyrometer consists of a hot body, lens, filament lamp, red filter placed in one line.
- Lens is used to pick up the radiant energy and to focus it on to the reference filament (lamp or bulb). In short, an image of radiating source is produced by lens and made to coincide with reference filament (lamp or bulb).
- Current through the reference filament (lamp or bulb) is varied to adjust the lamp intensity. The filament is viewed through an eyepiece and a red filter.

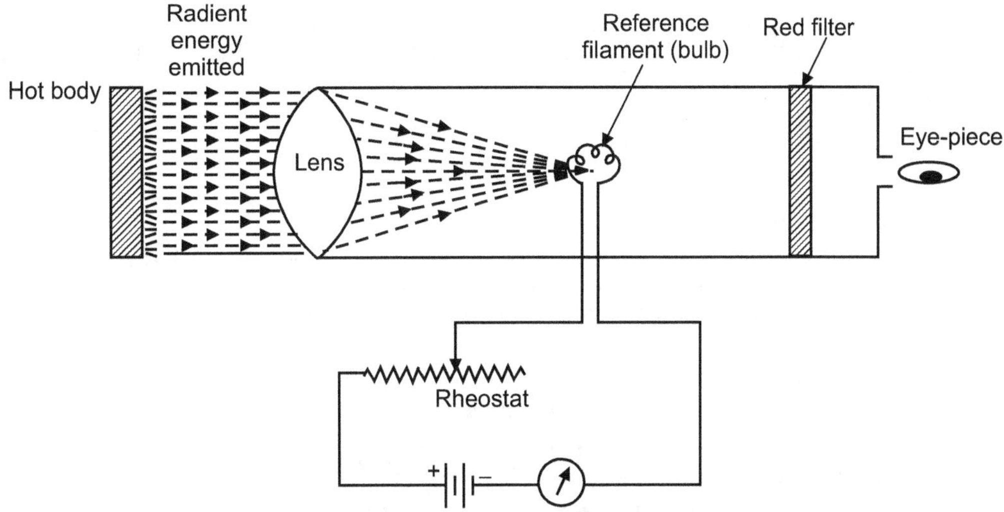

Fig. 2.31: Optical Pyrometer

- Current through the filament is adjusted until the filament image and image of radiating source are of equal brightness.
- This amount of current adjustment needed to match the images of two sources of light is done with the help of rheostat. Current adjustment is nothing but the measure of temperature.
- Position of rheostat indicated on the dial is graduated in terms of temperature.
- Red filter is placed in between heating filament and eye-position of operator. Red filter acts as a protective barrier for capturing unwanted radiations, which can lead to eye diseases of operator.

Conclusions:

1. When both, brightness of image produced by source and brightness produced by filament, are equal, the outline of filament will show a pattern as shown in Fig. 2.32 (a).
2. But, if temperature of filament is higher than that required for equal brightness, then filament becomes too bright and it can be seen in Fig. 2.32 (b).
3. If temperature of filament is lower than that required for equal brightness, then filament becomes too dark, which can be seen in Fig. 2.32 (c).

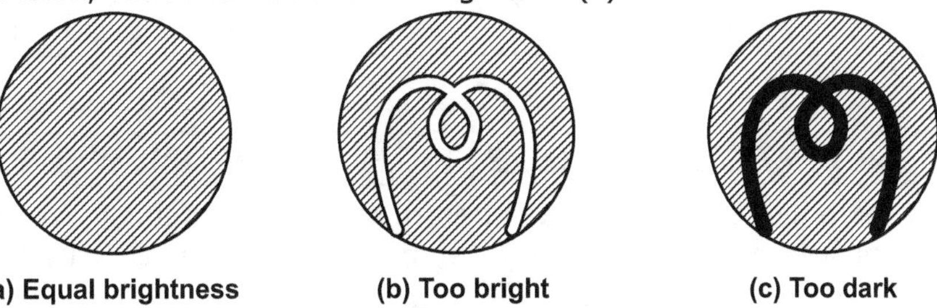

(a) Equal brightness   **(b) Too bright**   **(c) Too dark**

Fig. 2.32: Conclusions

2.47 ADVANTAGES, DISADVANTAGES AND APPLICATIONS OF OPTICAL PYROMETER

2.47.1 Advantages of Optical Pyrometer

1. Optical pyrometer is portable.
2. Fast response.
3. It can measure the high temperature with accuracy.
4. It is used to detect the temperature at distant place.
5. It is less sensitive to changes in emissivity.
6. It is used for very high temperatures (700°C to 3000°C).

2.47.2 Disadvantages of Optical Pyrometer

1. Device is very costly.
2. More chances of human error, while adjusting the image.
3. It measures the temperature of only hot surfaces.
4. Non-linear in behaviour.
5. Possible errors due to presence of intervening gases or vapours, which absorb radiations.

2.47.3 Applications of Optical Pyrometer

1. Temperature measurement in furnace, molten metal, etc.
2. Very high temperature applications of the order of 3000°C.

2.48 DIFFERENCE BETWEEN RADIATION PYROMETER AND OPTICAL PYROMETER

Sr. No.	Comparative Point	Radiation Pyrometer	Optical Pyrometer
1.	Temperature	More than 550°C.	More than 700°C and up to 3000°C.
2.	Sensitivity	Fair.	Good (Better).
3.	Calibration	By comparison with standard optical pyrometer.	Against standard tungsten strip lamp.
4.	Accuracy	More.	Less.
5.	Stability	Good (Better).	Fair.
6.	Working	Radiant energy is focused by lens on hot junction of thermocouple.	Colour of hot body is compared with heating filament.
7.	Output	Emf, which is calibrated to give temperature of hot body.	Colour, which is compared with heating filament, whose temperature is known.
8.	Cost	Low.	High.

2.49 IMPORTANT QUESTIONS WITH ANSWERS

Question 1: Select suitable system for temperature measurement in following applications. Justify answer with reasons. (i) Temperature in furnace, (ii) Engine temperature, (iii) Temperature in ovens.

Answer:

(i) Application	Suitable Measuring Instruments
Temperature in furnace	Thermocouple and Pyrometer

Reason: Furnace temperature ranges from 100°C to 3000°C.

(i) Thermocouples:

- **For temperature range of 100°C to 1500°C, thermocouples are used.**
- Various types of thermocouples are available depending upon the requirement.
- They are capable of measuring extremely wide range of temperature, from 100°C to 1500°C.

(ii) Pyrometers:

- **For temperature range of 1500°C to 3000°C, pyrometers are used.**

- Pyrometers are preferred in high temperature applications. This is a non-contact type of instrument.
- We can have its read-out scale at a considerable distance away from the applications for protection against high temperature.
- In such arrangement, where sensing element and read out device are located at considerable distance away from each other, temperature measured is accurate.

(ii) Application	Suitable Measuring Instruments
Engine temperature	Platinum RTD and Thermocouple

Reason: Engine temperature ranges from 500°C to 1200°C.

(i) **Platinum RTD** due to its high repeatability, linear characteristic, ability to take reading from distant place.

(ii) **Thermocouple**, because it gives high output with good stability and low cost.

(iii) Application	Suitable Measuring Instruments
Temperature in ovens	Pressure thermometer, RTD, Thermistors and Thermocouple

Reason: Variety of ovens are available having different temperature range (0°C to 300°C).

(i) **Pressure thermometer** due to lower cost, smaller bulb size, fast speed of response, most linear scale.

(ii) **RTD** due to high accuracy, wide temperature range from −200°C to 650°C, fast response, good reproducibility.

(iii) **Thermistors** due to small size and fast response, low cost, greater sensitivity.

(iv) **Thermocouple** due to its high output with good stability and low cost etc.

Question 2: State the thermometers most suitable for measurement of following temperatures,

(a) − 35°C to 510°C,

(b) − 65°C to 430°C,

(c) − 100°C to 315°C,

(d) − 15°C to 3870°C.

Answer:

Sr. No.	Temperature Range	Most Suitable Thermometer
1.	−35°C to 510°C	Mercury filled pressure thermometer.
2.	−65°C to 430°C	Bimetallic thermometer.
3.	−100°C to 315°C	Metallic oxide type thermistor or Pressure gauge thermometer.
4.	−15°C to 3870°C	Optical pyrometer.

2.50 SUMMARY OF TEMPERATURE MEASURING INSTRUMENTS

Type	Instrument	Working Principle	Application
Expansion thermometer	**Bimetallic thermometer**	Operation based on the principle of difference in the thermal coefficients of expansion of different metals.	Used in industries to measure temperature.
Filled system thermometer	**Liquid filled thermometer**	Based on the principle of volumetric expansion of fluid due to increase in temperature of fluid.	Used to measure temperature.

contd. ...

Electrical temperature measuring instruments	(i) Resistance Temperature Detector (RTD)	Resistance of pure metal wire with high positive coefficient of resistance increase with increase in temperature.	Measurement of temperature and radiant heat.
	(ii) Thermistor	Resistance of certain metallic oxides with negative temperature coefficient of resistance decreases with increase in temperature.	Measurement of thermal conductivity, vacuum and power at high frequency.
	(iii) Thermo-couple	An e.m.f. is generated across the cold junction of two dissimilar metals, when a hot body, whose temperature is to be measured, is placed near the hot junction of two dissimilar metals.	Temperature measurement of liquids and gases in storage tanks and flowing pipes, in industrial furnace, in cryogenic range.
Pyrometers	(i) Radiation pyrometer	Operation based on measurement of energy emitted by hot body.	Measurement of temperature above the operating range of thermocouple.
	(ii) Optical pyrometer	It is based on monochromatic radiations from hot body, whose temperature is to be measured.	For measurement of temperature of furnace, molten metal and other heated material.

Important Points

- **Working principle of capacitive transducer:** It works on the principle of change in capacitance.
- **Capacitive transducer** consists of two or more metal plate conductors separated by a dielectric material, which acts as an insulator. The dielectric material may be air, mica, oil, paper etc.
- **Linear potentiometer** is a device, which converts mechanical displacement into an electrical output. Linear potentiometer consists of a stretched resistance wire and a sliding or movable contact (wiper).
- **Rotational type potentiometer** is used for measurement for angular displacement.
- The resistive elements of potentiometer are metal alloys wound on a cylindrical ceramic former.
- **Wiper (slider)** is made up of phosphor bronze and it is gold plated for good electrical contact.
- **LVDT** is the most commonly used inductive transducer to translate linear motion into electrical signal (displacement). LVDT consists of an insulating hollow cylinder made up of Bakelite or any insulating material. On the circumference of this insulting cylinder, one primary winding (P) and two secondary windings (S_1 and S_2) are wound. Inside the hollow insulating cylinder, magnetic or armature core is placed, which is free to move back and forth.
- **RVDT** is used to sense the angular displacement. RVDT is also an inductive transducer, which converts angular displacement into an electrical signal.
- **Temperature** is defined as, "the degree of hotness and coldness of a body or an environment measured on definite scale." Also, Temperature can be defined as, "the driving force or potential causing flow of energy in the form of heat."
- **Temperature measuring devices** working on electrical methods of temperature measurement are called as Electrical transducers. These electrical transducers convert the measured temperature into an equivalent electrical signal, such as, resistance in RTD & thermistor, whereas, voltage in thermocouple.

- **Working principle of resistance thermometer (RTD):** "Electric resistance of metals is directly proportional to temperature, i.e. electric resistance of metals increases with increase in temperature."
- The term **"thermistor"** means Thermal – Resistor. It is also called as, **"Temperature Sensitive Semiconductor".**
- **Working principle of thermistor:** Thermistors work on following principles.
  - (a) Negative temperature coefficient (NTC), where resistance of material decreases with increase in temperature. Relationship of temperature versus resistance is non-linear, i.e. exponential. Materials possessing the property "NTC" are composed of metal oxides such as, Manganese, Nickel, Cobalt, Copper, Iron and Uranium.
  - (b) Positive temperature coefficient (PTC), where resistance of material increases with increase in temperature. Materials possessing the property "PTC" are composed of metal oxides, such as, Powdered Barium Carbonate, Barium Titanate and Titanium Oxide.
- **Working principle of thermocouple:** If temperature at one junction (reference junction, i.e. cold junction) is kept constant, then effective emf will be generated due to change in temperature at another junction, i.e. hot junction. Temperature change can be measured by this effective emf generated in thermocouple.
- **Seebeck effect** states that, "when two dissimilar metal wires are joined to form a complete electric circuit and the two junctions are maintained at different temperatures, an electromotive force (emf) is set up to establish flow of current."
- **Thomson effect** gives the emf generated or developed between two ends of a single wire, when one end is heated. Two emfs will be generated, since there are two wires.
- **Peltier effect** gives the emf generated due to contact of two different metals. Two emfs will be generated, since there are two junctions. In thermocouple, the emf measured is equal to Net emf, which is resultant of two Peltier emfs and two Thomson emfs.
- **Thermo-electric laws:**
  - **(i)** "Application of heat to single homogeneous metal is not capable of producing or sustaining electric current."
  - **(ii) Law of intermediate metals:** "Introduction of third metal into a thermocouple circuit will have no effect on the emf generated, as long as, the junctions of third metal with the thermocouple metals are at same temperature."
  - **(iii) Law of intermediate temperature:** If a thermocouple circuit develops an emf e_1, when its junctions are at temperature T_1 and T_2 and an emf e_2, when its junctions are at temperature T_2 and T_3, then it will develop an emf equal to $(e_1 + e_2)$, when its junctions are at temperature T_1 and T_3."
- **Non-electrical methods of temperature measurement** are, Bimetallic thermometers, Liquid in glass thermometer, Liquid in metal thermometer, Vapour pressure thermometer etc.
- **Fundamental principles** used by **Bimetallic thermometers** are,
  - (a) All metals expand or contract with change in temperature.
  - (b) Temperature coefficients of expansion are not same for all metals and therefore their rates of expansion or contraction are different. This property of difference in rates of thermal expansion is used to produce deflections proportional to temperature changes.
- **Working principle of liquid in glass type thermometer:** If temperature increases, liquid expands in volume. This change in volume of liquid is directly proportional to change in temperature.
- **Working principle of "Liquid in metal thermometer" or "Pressure gauge thermometer":** If temperature increases, liquid expands in volume. This increase in volume of liquid is directly proportional to increase in temperature. Therefore, by measuring increase in volume of liquid, we can carry out temperature measurements.
- **Working principle of vapour pressure thermometer:** "Vapour pressure of a liquid depends upon the temperature. If this liquid is partially filled in an enclosed chamber, it will create a definite vapour pressure, by which, temperature can be measured".
- **Pyrometry** is contact-less method of temperature measurement. It is the technique of temperature measurement of a body by measuring its electromagnetic radiations.
- **Pyrometer** is used, when temperature being measured is very high and physical contact is impracticable or impossible.

- **Working principle of pyrometer:** When a body is heated, it emits thermal energy in the form of electromagnetic radiations. This property is used as working principle of pyrometer.
- **Working principle of radiation pyrometer:** "Temperature of a body can be measured by measuring radiant energy emitted by that hot body."
- **Working principle of optical pyrometer:** It is based on comparisons of intensity (brightness) of visual radiation emitted by a hot body with the radiation energy emitted by a source of known intensity. Here, the intensity of radiation emitted by hot body, whose temperature is to be measured, is matched with the intensity of a reference filament (lamp), whose temperature is known.

Theory Questions for Practice

1. List advantages and disadvantages of capacitive transducer used for measurement of displacement.
2. Explain the use of wire wound potentiometers for the measurement of linear and rotary motions.
3. State advantages and disadvantages of potentiometer.
4. State application of potentiometer and write its working principle.
5. State any four applications of linear potentiometer.
6. Explain with neat sketch, working principle of LVDT.
7. Explain the construction and working of LVDT.
8. With the help of labelled sketch, explain Displacement Measurement by LVDT.
9. Explain how LVDT is used for measurement of displacement.
10. What are the characteristics of LVDT?
11. Draw neat sketch of RVDT. Explain its working and state any two applications.
12. State specifications of RVDT.
13. Write the specification for displacement transducer.
14. Draw neat constructional details of resistance thermometer and explain its working.
15. Explain with neat sketch, liquid in glass thermometer.
16. Draw neat sketch of 'Radiation pyrometer' and explain how it is used for measurement of temperature.
17. Draw neat sketch of optical pyrometer and explain its working.
18. State laws of 'Intermediate temperature' and 'Intermediate metal' with neat sketches.
19. Explain 'Seebeck effect' and 'Peltier effect'.
20. Differentiate between RTD and thermistor.
21. Write any four advantages of liquid in glass thermometer method.
22. Explain the working principle of platinum resistance thermometer.
23. Define thermocouple? State the materials of thermocouple.
24. List the electrical and non-electrical method for temperature measurement. Explain with neat sketch, liquid in glass thermometer.
25. Define thermistor? Explain the working of a thermistor.
26. Explain the Seebeck and Peltier effect.
27. Classify the temperature measuring instruments and indicate approximate temperature range of each category.
28. Explain working of platinum resistance thermometer with a neat sketch.
29. Explain with neat sketch, working of any one thermo-resistance transducer.
30. Explain with neat sketch, the construction and working of thermocouple.
31. Enlist the different materials used for developing thermocouple?
32. Define thermistors? Explain principle of operation and give advantages.
33. Compare between thermocouple and thermistor on the basis of: Construction, Temperature Range, Size, Sensitivity, Cost, and Effect of ambient condition.
34. Which instrument is used for measuring temperature of furnace having temperature 1400°C? Justify your answer.
35. Explain construction and working of bimetallic thermometer.
36. Explain working principle of RTD? Explain with necessary sketch.

■■■

FLOW MEASUREMENTS

3.1 INTRODUCTION TO FLUID FLOW

- Whenever a fluid is in motion, large number of small fluid particles are grouped together to form the flowing stream.
- These particles, while moving, group themselves in a variety of ways. The type of flow may be according to the conditions, under which, the particles of water unite and move.
- **Flow may occur due to:**
  (a) Difference of energy between two distant points.
  (b) Due to pressure difference between two ends of pipe (horizontal).
  (c) Due to difference in datum energy and pressure energy in case of inclined pipes.

3.2 FLUID FLOW MEASUREMENTS

- **Flow measurement** is defined as, *"the process of measuring amount of fluid flowing through a given point in a given (definite) time period."*
- After measuring amount of fluid and time taken to flow, we get,

$$\text{Average flow rate, } Q = \frac{\text{Total amount of flow measured}}{\text{Time taken to flow}}$$

- **Rate of flow** or **Discharge** is defined as, *"the quantity of a fluid, flowing per second through a section of a pipe."*

3.3 FLOW METER

- Flow meter is capable of measuring quantity of fluid flowing in a circular pipe.

3.3.1 Classification of Flow Meters

Flow measuring meters are classified as,

1. **Rate meters (Inferential):**
   - The term **"Inferential"** is applied to those flow-meters, which do not measure the flow directly. Such type of flow meters measure other quantities, such as, pressure, temperature etc., which are related to flow. Therefore, they are called as **"Inferential type flow meters"**.
   - Primary sensing element of inferential type flow meters gives an indication, which is converted into velocity of flow. This velocity of flow (in conjunction with cross-sectional area "A" of the flow meter) gives the discharge (volume flow rate) by using the relation, $Q = A \times V$.
   - **Rate meters are further classified as,**
     (i) **Constant Area Variable Head Meters** such as Venturimeter, Flow nozzle, Orifice meter, Pitot tube etc.
     (ii) **Constant Head Variable Area Meters** such as Rotameter, Cylinder and Piston arrangement.
     (iii) **Variable Velocity Meters** such as Hot wire anemometers, Current meters.

2. Quantity meters (Positive displacement meters):

- In this method, primary sensing element of meter divides the flow of fluid into isolated quantities.
- Then, each quantity is separately measured either by weight or by volume.
- Secondary element of meter is called as **counter**, which acts as a **totalizer**.
- Instruments belonging to this category are called as **positive displacement meters**.
- **Examples:**
  - (i)   Coriolis Flow Meter,
  - (ii) Oscillating Piston Flow Meter,
  - (iii) Rotating Vane Flow Meter etc.

3. Special meters:

- (i)   Ultrasonic Flow Meter,
- (ii) Electromagnetic Flow Meter,
- (iii) Turbine Meter.

3.4 VARIABLE HEAD OR DIFFERENTIAL FLOW METER

- Variable head meters operate on the principle that, *"if a restriction element (obstruction) such as orifice plate or venturi-tube or elbow in the pipe of flowing fluid is introduced, then it produces a pressure difference across the restriction element used."*
- This pressure difference is proportional to the flow rate.
- Therefore, such flow meters are called as **restriction type flow meters**, which results in acceleration of fluid causing change (decrease) in pressure.
- In other words, pressure drops, because, some pressure energy of fluid is converted into kinetic energy, which increases velocity of fluid in downstream side.
- This magnitude of pressure drop or change in pressure can be measured.
- Since flow is directly proportional to change in pressure, therefore, we say that, pressure difference can be used as function of flow to find its value.

  Mathematically,   Flow $\propto \sqrt{\delta P}$

- Change in pressure before and after the obstruction (i.e pressure difference between upstream side and downstream side) is measured by a U-tube differential manometer.

3.4.1 Advantages of Variable Head Meters

1. Low cost.
2. It has wide range of application.
3. Accuracy up to $\pm$ 2%.
4. It can be adopted for any size of pipe and any flow rate.
5. It can be easily removed without shutting down the process.

3.4.2 Disadvantages of Variable Head Meters

1. Difficult to measure flow rate of slurry.
2. High permanent pressure loss.
3. It does not have linear characteristic.
4. Low flow rate is not easily measured.

3.4.3 Difference between Variable Head Type Flow Meter and Variable Area Type Flow Meter with Constant Head

Variable Head Meters	Variable Area Meters (Constant Head)
• In case of variable head meters, pressure difference is used to measure flow rate. In these meters, cross-sectional area is reduced at the point of measurement.	• Working principle of variable area meter is exactly reverse of variable head meters.
• In case of variable head flow meters, the pressure difference across the two sections is the measure of flow rate.	• In variable area meter, the differential pressure between top face and bottom face of float is maintained constant, and variation in cross-sectional area of the flow is measure of flow rate.
• In such cases, the flow restriction is of fixed size and the pressure difference across the two section changes with flow rate.	• Here, the size of restriction is adjusted by an amount, which is proportional to flow rate.
• **For example: Venturimeter, Orifice plate, Flow nozzle, Pitot tube.**	• Such types of flow meters are called as constant head and variable area type flow meters.
	• If differential pressure head is maintained constant by adjusting out flow area A_2 of a construction meter, then, the outflow area at any instant would be a measure of flow rate through the device.
	• **For example: Rotameter**

3.5 VARIABLE HEAD TYPE FLOW METERS

- Following are the different types of variable head type flow meters.
  1. Venture tube or Venturimeter,
  2. Orifice plate or orifice meter,
  3. Pitot tube.

3.6 VENTURE TUBE OR VENTURIMETER

Applications:
- Venturimeter is a device used to measure rate of discharge in a pipeline.
- It is fixed permanently at different sections of pipeline to know the discharges at those sections.

Construction:
Venturimeter consists of following three parts:

(a) Convergent cone:
- It is a short pipe, which converges from a diameter d_1 (diameter of pipe, in which, Venturimeter is fitted) to a smaller diameter d_2 (diameter of throat).
- The throat ratio $\frac{d_2}{d_1}$ varies between $\frac{1}{4}$ or $\frac{1}{5}$. In other words, the slope of converging cone is between 1 in 4 or 1 in 5.

(b) Throat:
- It is a short pipe having diameter d_2, connected between converging and diverging cones.
- Diameter of throat varies between 33% − 75% of main pipe diameter. Preferably, it is taken as 0.5 times the diameter of pipe. i.e. $d_2 = 0.5\ d_1$.

(c) Divergent cone:
- It is a long pipe, which diverges from a diameter d_2 to original pipe diameter d_1.

Working:
- One end of U-tube differential manometer is connected to the pipe at entrance section (before converging section) and another end is connected at the throat. This differential manometer is used for the measurement of pressure difference.

- When the liquid flows through the venturimeter (converging cone), its flow is accelerated resulting in an increase in velocity. Therefore, the velocity of liquid at section 2 (throat) becomes higher than that at section 1. This decreases the pressure at section 2.
- When the liquid flows through divergent cone, its flow is decelerated (Retarded). As a result of this retardation, the velocity of liquid decreases with subsequent increase in pressure.
- Thus, venturimeter works on the principle of converting pressure energy (head) into kinetic energy by reducing cross-sectional area of flow passage. This leads to pressure difference, which is measured by means of U-tube differential manometer.

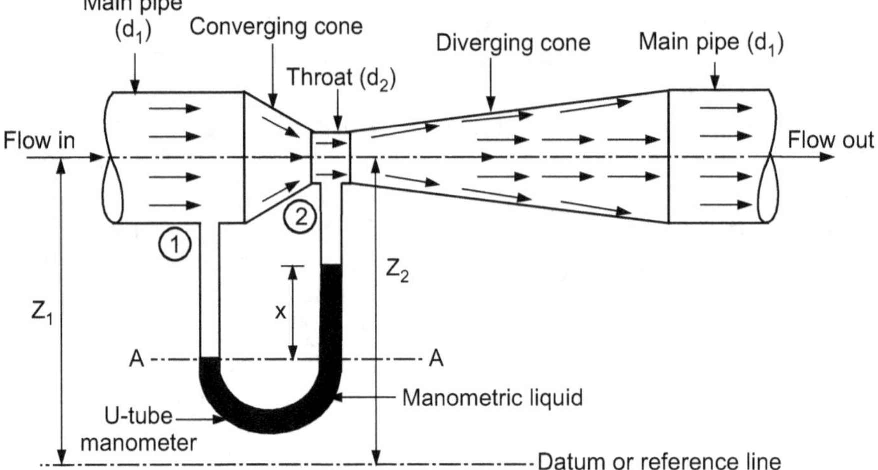

Fig. 3.1: Venturimeter

- Flow rate of fluid can be measured using the equation/ formula given below:

$$Q_{act} = \frac{C_d\, A_1 A_2}{\sqrt{A_1^2 - A_2^2}} \cdot \sqrt{2gh} \qquad \text{in 'm}^3/\text{sec'}$$

Where,
C_d = Coefficient of discharge = 0.98
h = Pressure difference   in 'm'
A_1 = Area of pipe   in 'm²'
A_2 = Area of throat   in 'm²'

3.6.1 Advantages of Venture Tube

1. It causes low permanent pressure loss.
2. It is widely used for high flow rates.
3. It is available in very large pipe sizes.
4. It is more accurate over wide flow range than orifice plates.

3.6.2 Disadvantages of Venture Tube

1. High cost.
2. It is not useful for pipe diameters less than 76.2 mm.
3. Difficult to inspect due to complex construction.

3.6.3 Applications of Venture Tube

- For flow measurement of,
  1. Water.
  2. Fluid used in industrial processes.
  3. Industrial wastes.
  4. Slurries and dirty liquids.
  5. Fluids with suspended particles.

3.7 ORIFICE OR ORIFICE METER

Application:

- Orifice is used in orifice meter, which measures discharge of fluid flowing through the pipe.
- Here, orifice (i.e. orifice plate) is sharp edge circular opening having hole diameter as 0.4 to 0.8 times diameter of pipe. Preferably, it is taken 0.5 times the diameter of pipe.

Construction and Working:

- Orifice is made up of materials like stainless steel, phosphor bronze etc., to withstand against corrosive effect of flowing fluids.
- A U-tube differential mercury manometer is connected between two sections, one on upstream side and other at vena contracta on the downstream side of orifice plate.
- Orifice meter works on same principle as that of Venturimeter i.e. the pressure head (energy) is converted into kinetic energy in an accelerated flow.
- Thus, orifice meter works on the principle of converting pressure energy (head) into kinetic energy by reducing cross-sectional area of flow passage. This leads to pressure difference, which is measured by means of U-tube differential manometer.

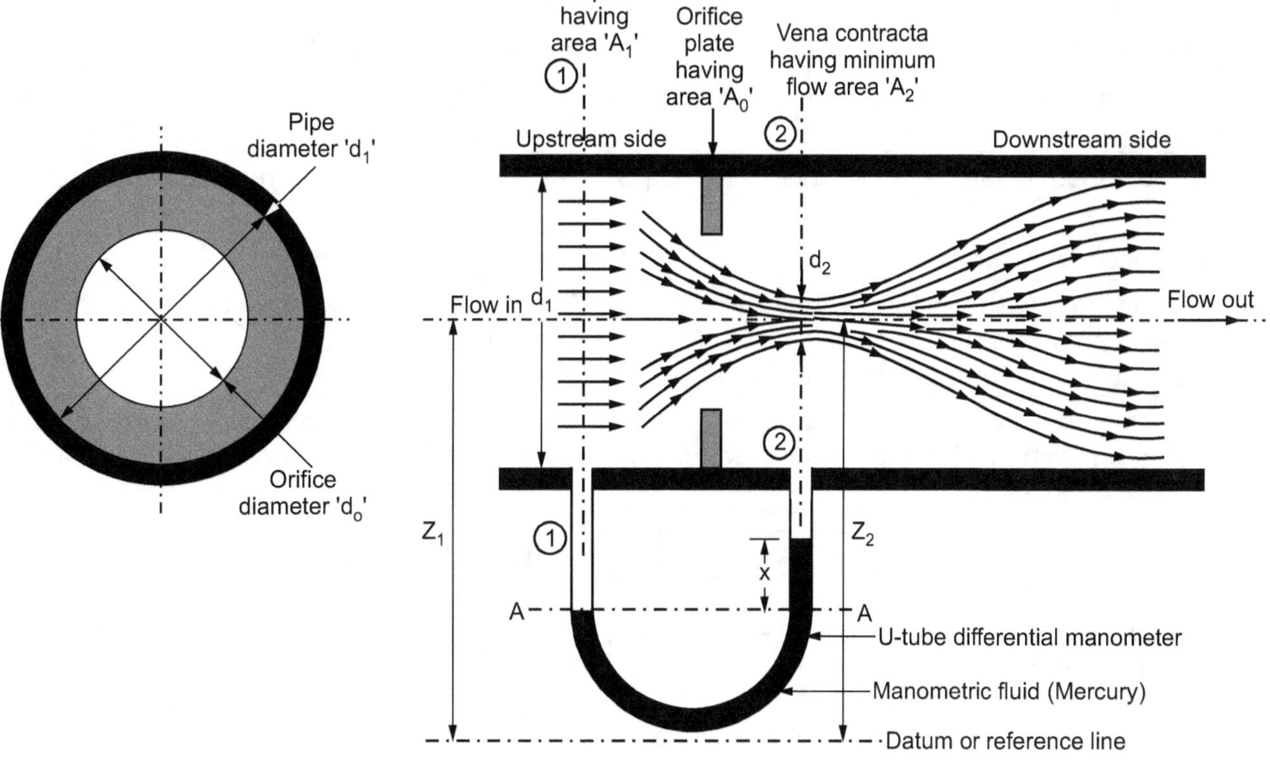

Fig. 3.2: Orifice Meter

- Flow rate of fluid can be measured using the equation/ formula given below.

$$Q_{act} = \frac{C_d\, A_1 A_2}{\sqrt{A_1^2 - A_2^2}} \cdot \sqrt{2gh} \qquad \text{in 'm}^3\text{/sec'}$$

Where,

C_d = Coefficient of discharge = 0.64

h = Pressure difference   in 'm'

A_1 = Area of pipe   in 'm²'

A_2 = Area of Vena-contracta   in 'm²'

3.7.1 Advantages of Orifice Meter

1. Low cost.
2. It can be used with differential pressure devices.
3. It is available in many materials.
4. It has well known and predictable characteristics.

3.7.2 Disadvantages of Orifice Meter

1. It causes relatively high permanent pressure loss.
2. It tends to clog. Therefore, its use is restricted in slurry applications.
3. Its accuracy depends upon the care taken during its installation.
4. It is subjected to erosion, corrosion and scaling.

3.7.3 Applications of Orifice Meter

1. Widely used in the applications, where space availability is a problem.
2. Useful for wide range of pipe diameters ranging from 15 mm to 1.5 m diameter pipes.

3.7.4 Difference between Venturimeter and Orifice Meter

Venturimeter	Orifice meter
1. Losses are very low and hence, coefficient of discharge (C_d) is high. In general, C_d = 0.98.	1. Losses are high and hence, coefficient of discharge (C_d) is less. In general, C_d = 0.64
2. Suitable for measurements of higher flow rates.	2. Not suitable for measurements of higher flow rates.
3. Cost of installation and replacement is more.	3. Cost of installation and replacement is less.
4. More space is required.	4. Less space is required.
5. They can be used to measure flow rates of all incompressible fluids (gases with low variations in pressure as well as liquids).	5. They are generally used to measure flow rates of liquids.
6. Here, the flow is accelerated gradually in converging section with subsequent gradual decrease in pressure.	6. Here, the entire pressure drop occurs instantly, when flow passes through orifice.

3.9 PITOT TUBE

Application:
- Pitot tube is used to measure the velocity of fluid flowing through a pipe at any point.

Working Principle:
- It works on the principle that, *"if velocity of flow at a point becomes zero, there is an increase in pressure energy"*. This point is known as *stagnation point*.

Construction:
- Pitot tube consists of a glass tube opened at both the ends.
- The lower end is bent through 90° and is placed such that, it faces the flow of liquid coming from upstream direction.
- The liquid enters into this lower end and then rises up vertically in the straight end of tube due to the conversion of kinetic energy into pressure energy.
- The velocity of liquid (V) is determined by measuring the rise of liquid in the tube.

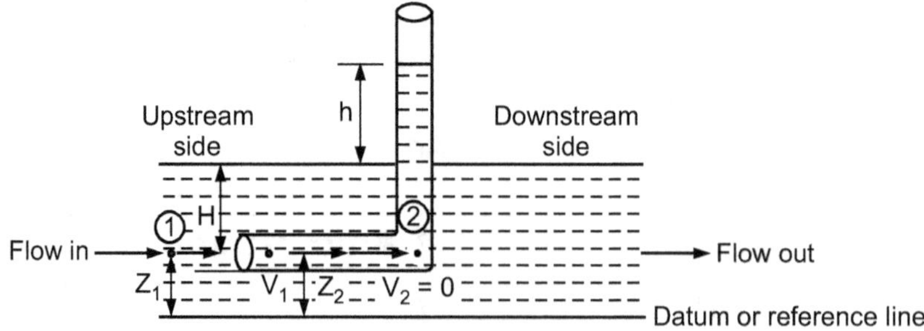

Fig. 3.3: Pitot Tube

- Flow rate of liquid can be determined by,
$$Q = A \cdot V$$

3.9.1 Advantages of Pitot Tube

1. No pressure loss.
2. Economical to install.
3. It can be easily removed from pipe line.

3.9.2 Disadvantages of Pitot Tube

1. Poor accuracy.
2. Unsuitable for dirty or sticky fluids.
3. Highly sensitive to upstream disturbances.

3.10 FLOW NOZZLE METER

- A flow nozzle operates on the principle of variable head type flow meter.
- Differential pressure is developed by inserting flow nozzle in the pipe as shown in Fig. 3.4.
- A flow nozzle consists of a convergent inlet, whose shape is quarter ellipse with cylindrical throat. Differential pressure measurement taps are provided on pipe diameter upstream and one-half diameter downstream from the inlet faces of nozzle.
- Stainless steel is the most commonly used material for manufacturing of flow nozzle.
- Discharge or the flow rate depends upon the pressure difference measured by U-tube differential manometer.
- Discharge can be calculated by using the formulae used in case of venturimeter and orifice meter discussed in previous sub-articles.

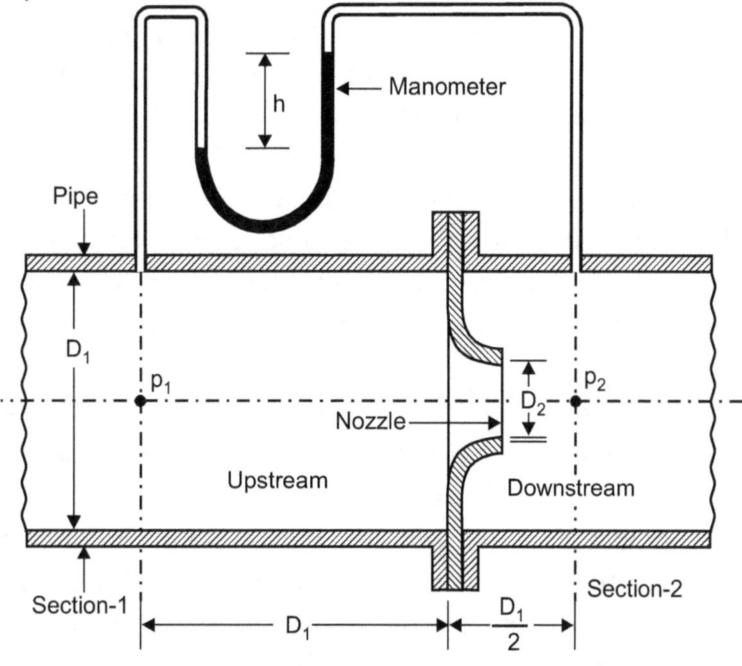

Fig. 3.4: Flow Nozzle

3.10.1 Advantages of Flow Nozzle

1. They have higher coefficient of discharge than that of orifice meter.
2. Cheaper and easier to install.
3. Physical length is less as compared to venturimeter.
4. Quite useful for measurement of flow of fluids containing settled solids.

5. Widely used for high pressure/temperature steam flow.

6. Being more rugged and more resistant to erosion than sharp edged orifice, they can be used for flow measurements at high velocities.

3.10.2 Disadvantages of Flow Nozzle

1. Owing to poor pressure recovery, they are not useful for applications, where pressure heads are small or where pressure recovery is a must.

2. It is expensive and difficult to install as compared to orifice meter.

3. Limited moderate pipe sizes, not available above 120 cm.

4. Higher maintenance.

3.11 ROTAMETER (VARIABLE AREA TYPE FLOW METER)

- Rotameter is most commonly used type of **variable area type flow meter**.
- Here, flow area is varied by means of float kept in a tapered glass tube.
- Depending on the rate of flow, the float takes position in the tube, which increases or decreases size of area, thus keeping the pressure difference constant.

Principle of Working:

- *"Pressure difference across an annular orifice is directly proportional to square of its flow area and square of flow rate."*

Construction:

- Rotameter consists of a transparent and tapered glass tube placed vertically in such a way that, the larger end is at top.
- A float is provided inside the glass tube, which is free to move up and down.
- Shape of float is selected, in such a way that, it creates minimal obstruction for flow.
- Density of float is more (nearly double) than the density of the fluid, whose flow rate is to be measured.
- Glass tube is made tapered to maintain constant pressure drop. This arrangement gives a linear relationship between the flow rate and position of float within the tube.
- Flow inlet is provided at bottom of the glass tube and flow outlet is at the top of tube.
- Under equilibrium, fluid force acting vertically upward is equal to weight of float acting vertically downward.

$$\text{Weight of float} = \text{Mass of float} \times \text{Gravitational acceleration} = m \times g$$
$$\text{Fluid force} = \text{Pressure of fluid} \times \text{Internal Cross-sectional area of Rotameter}$$
$$= P \cdot A$$

- A linear scale is graduated or marked on the outer periphery of glass tube.
- An annular orifice is the free area between the float and inside wall of tube.
- Float adjusts its position according to discharge through the passage, i.e. float rises higher or lower depending on the flow rate.

Working:

- Fluid (whose flow is to be measured) enters the tube from bottom, then moves upward and escapes through the top.
- When there is no flow through rotameter, the float will rest at the bottom of glass tube.
- In this position, the maximum diameter of float is approximately same as the inside diameter of glass tube.

- When the fluid enters in the metering tube, the float moves upwards, increasing the flow area of **annular orifice.**

- Float is pushed upwards, until, the lifting force produced due to pressure difference across its upper and lower surface becomes equal to weight of float.

- Now, if the flow rate increases, the pressure difference and hence, lifting force increases temporarily. Due to this, float moves upward and increases the area of annular orifice, so that, the increased lifting force will decrease and fluid force will be equal to weight of float.

- Thus, the pressure difference remains constant by changing the area of annular orifice in proportion to the flow rate.

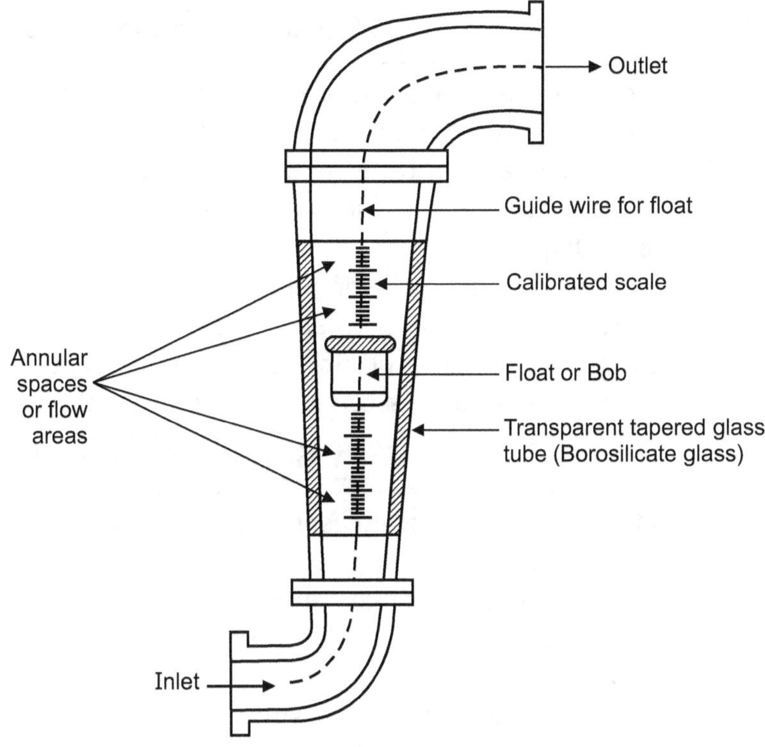

Fig. 3.5: Rotameter

- Every float position corresponds to one particular flow rate for a fluid of given density and viscosity.
- Calibrated scale marked on the glass tube provides the direct indication of flow rate.

Material used for float: Heavier metals like Stainless steel, Gun metal, Brass etc.

Material used for tube: Non-corrosive materials such as Glass, Plastic etc.

3.11.1 Characteristics of Rotameter

1. Simplicity of construction and low cost.

2. Possibility of remote indication and record.

3. Essentially a linear scale with most of the meters.

4. Easy to install.

5. Accuracy within ± 2% of the maximum reading.

3.11.2 Advantages of Rotameter

1. Pressure loss is nearly constant and small.

2. It can handle any corrosive fluid.

3. Suitable for measurement of gases as well as liquids.

4. Good accuracy at low flow rates.

5. Provides linear scale.

6. Condition of flow is easily visible.

7. Relatively low cost.

8. Good range of measurement.

9. Gives direct visual indication of flow rate.

10. Minimum piping required.

11. Easily equipped with data transmission, indicating and recording devices.

3.11.3 Disadvantages of Rotameter

1. It must be installed in vertical position only.
2. It becomes expensive in case of high pressure and high temperature applications.
3. When opaque (Non-transparent) fluid is used, float may not be visible.
4. It is not rigid in construction as compared to venturimeter or orifice meter, due to use of glass tube.
5. Not good to use in pulsating flow cases (i.e. fluctuating flow rates).
6. Limited to small pipe sizes and capacities.
7. Less accurate than venturimeter and orifice meter.
8. Not suitable for liquids carrying suspended solid particles.

3.11.4 Applications of Rotameter

- Rotameter is used for measurement of flow of liquids and gases in,
  1. Laboratories.
  2. Testing and production lines.
  3. Process industries.
  4. Oil industries.
  5. Oxygen flow rate measurement in medical applications.
  6. For metering purge flows (i.e. viscous fluids having suspended particles).

3.12 HOT WIRE ANEMOMETERS (VARIABLE VELOCITY TYPE FLOW METER)

- Hot wire anemometer is a type of variable velocity meter and uses thermal method for flow measurement.
- It uses the property that, resistance of wire is directly proportional to its temperature.

Working Principle:

- "When fluid flows over a heated surface, the surface temperature reduces due to transfer of heat from the heated surface to the fluid. This rate of reduction in temperature is a measure of flow rate."

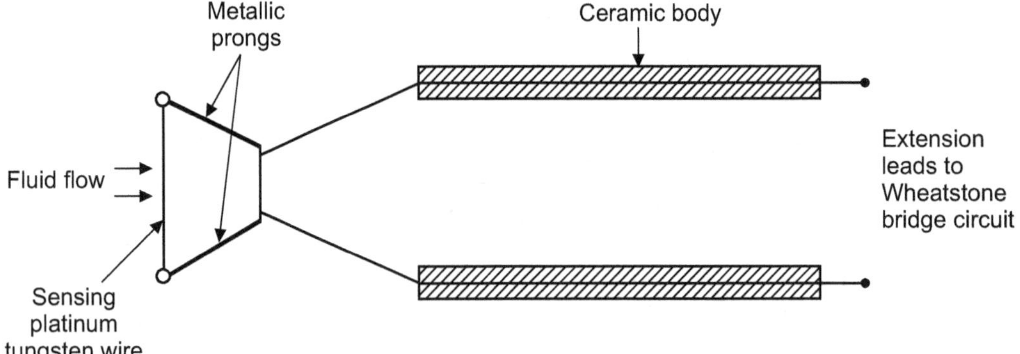

Fig. 3.6: Working Principle of Hot Wire Anemometer

Construction:

- Hot wire anemometer consists of a platinum-tungsten wire (with diameter 5 μm) of short length and stretched between two supports. This wire acts as a sensor.
- The stream of air or gas is kept perpendicular to wire.
- This wire of platinum-tungsten (sensor) is welded between the two prongs of probe.
- The wire (sensor) is heated electrically and forms one arm of the **Wheatstone's** bridge.
- The Wheatstone's bridge has a balance detecting galvanometer.

Working:

- Here, the wire is heated electrically. The temperature of wire is determined by measuring its resistance with Wheatstone's bridge.

- When the wire along with the probe is introduced in the flowing fluid, it gets cooled due to velocity of fluid and resistance of wire decreases.

- Rate of cooling of wire (sensor) depends upon following properties:

  1. Dimensions and physical properties of wire.

  2. Temperature difference between the wire and the fluid.

  3. Thermodynamic properties of fluid (i.e. P, V and T).

  4. Stream velocity of fluid flow under measurement (i.e. Velocity of fluid flow).

- Amongst all the above properties, fluid velocity is the variable and rest of the properties almost remain constant.

- Therefore, Stream velocity is the measure of instrument response or we can say that, **the instrument response is direct measure of flow velocity.**

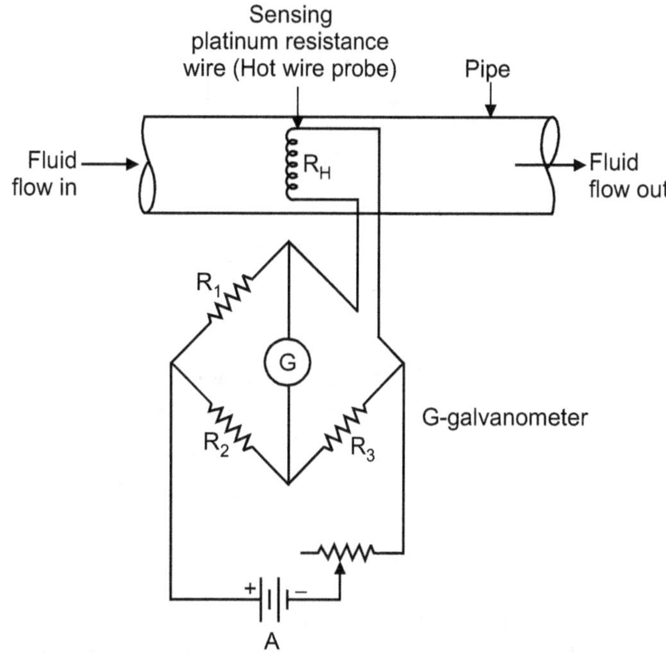

Fig. 3.7: Wheatstone's Bridge Circuit Arrangement for Flow Measurement of Fluid

Methods of Measurement:

- There are two methods of measuring fluid flow:

1. Constant current mode:

- Here, voltage across the bridge is kept constant i.e. heating current is maintained constant.

- Initially, the circuit is so adjusted that, when the heated wire lies in stationary air, the galvanometer reads zero.

- When the air flows, the hot wire (sensor) gets cooled and resistance of wire decreases causing the galvanometer to deflect.

- The galvanometer deflection is amplified, measured and correlated with the air velocity using a suitable calibration method.

2. Constant temperature mode:

- Here, the operating resistance of wire and hence, the temperature is maintained constant.

- When the air flows, the hot wire gets cooled due to the flowing fluid.

- To keep the operating resistance of wire and its temperature constant, external bridge voltage is applied to the wire.

- This external bridge voltage is varied and adjusted to bring the galvanometer pointer to zero position. The external bridge voltage supplied can be measured by voltmeter.

- Now, the reading of voltmeter is recorded and then correlated with air velocity using a suitable calibration method.

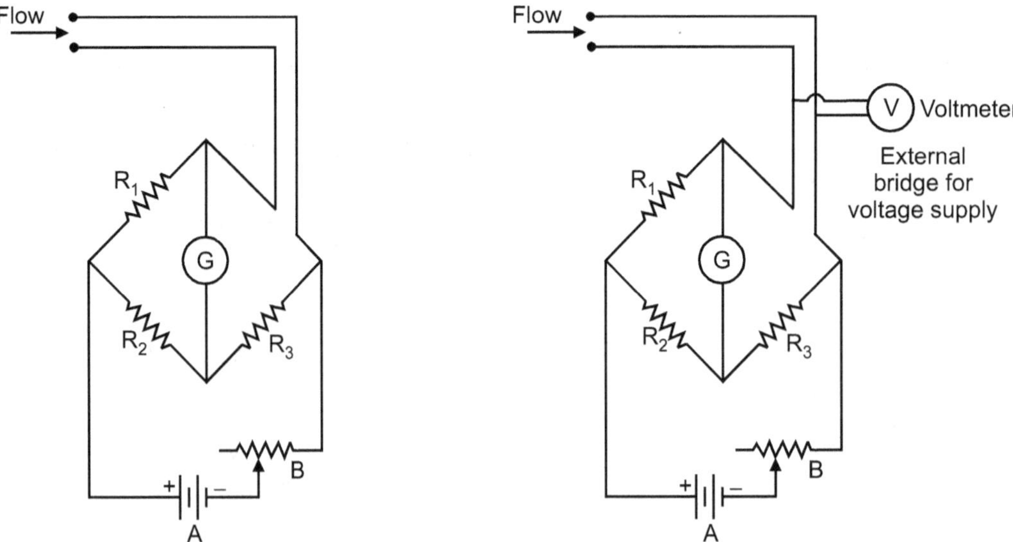

(a) Constant current mode **(b) Constant temperature mode**

Fig. 3.8: Methods of Fluid Flow Measurement by Hot Wire Anemometer

3.12.1 Advantages of Hot Wire Anemometer

1. Simple construction.
2. Direct relation of flow rate with current and temperature, which makes easier to calibrate instrument.
3. High mechanical strength.
4. It can be preferably used for flow measurement of fluctuating flow.
5. Output is electrical; therefore, the readout can be analog or digital.
6. Excellent accuracy up to ± 0.1 %.
7. Suitable for both, liquid and gas measurements.
8. Measurement in any direction is possible.

3.12.2 Disadvantages of Hot Wire Anemometer

1. It requires external power.
2. It can be used for clean fluid only.
3. Accumulation of dust particles on the wire leads to serious heat transfer error.
4. In constant current method, current has to be kept large with sudden drop in fluid velocity. Otherwise, there may be chance of lead getting burn out.
5. Fine wire (small / minute diameter) has limited physical strength.

3.12.3 Applications of Hot Wire Anemometer

1. It is widely used in measurement of air velocity, which is fluctuating in nature.
2. It is used in measurement of wind velocity.
3. In research applications to study varying flow conditions.
4. Used for flow measurements of liquid as well as gas.

3.13 ELECTROMAGNETIC FLOW METER

Working principle:

- Electromagnetic flowmeter works on the principle of **Faraday's law of electromagnetic induction** for flow measurement, according to which, *"Whenever conductor moves through magnetic field of given field strength, a voltage (e.m.f.) is induced in the conductor, which is proportional to relative velocity between the conductor and magnetic field."* **OR** *"Whenever a conductor cuts magnetic flux lines (magnetic field), an emf is induced in the conductor, which is directly proportional to rate of change of magnetic flux."*
- In case of electromagnetic flow meter, **fluid flow acts as a conductor.**

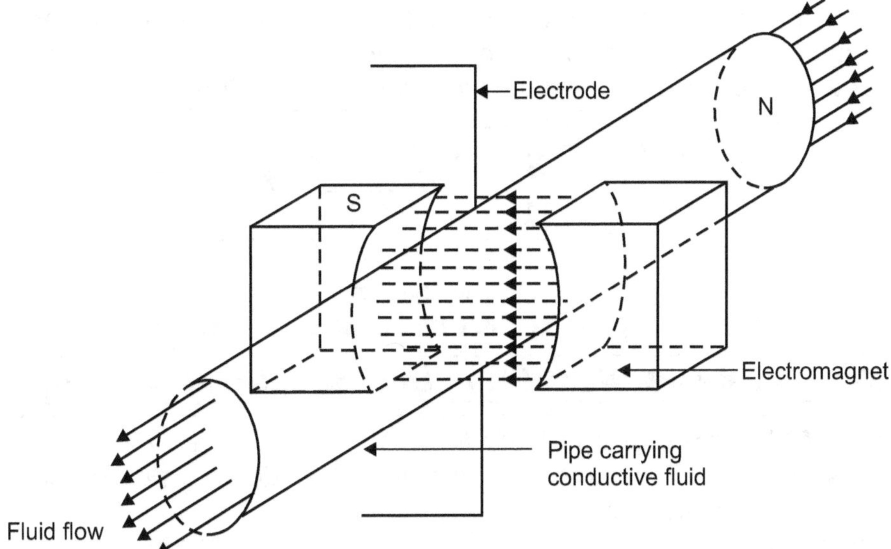

Fig. 3.9: Working Principle of Electromagnetic Flow Meter

Construction:

- Electromagnetic flow meter consists of electrically insulated or non-conducting pipe such as fiber glass.
- A pair of electrodes is mounted opposite to each other and flushes with the inside wall of pipe carrying the fluid, whose flow is to be measured.
- In the figure, it can be observed that, two electrodes are placed at right angles to the plane of magnetic field, i.e. magnetic flux lines.
- The pipe is surrounded by an electromagnet, which produces magnetic field.
- This magnetic field is generated by the current flowing through the coil wounded on the electromagnet. The coil is powered by a steady D.C. supply.

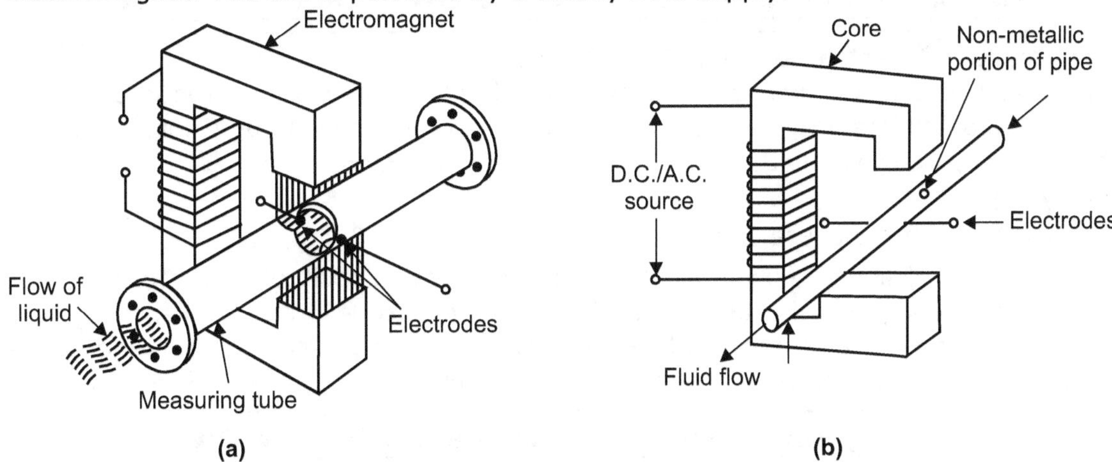

Fig. 3.10: Schematic Arrangement of Electromagnetic Flow Meter

Working:

- A conductive fluid is passed through the pipe.
- As the fluid passes, its motion relative to magnetic field produces an e.m.f proportional to velocity of fluid.
- It is given by Faraday's law as,

$$E = B.L.V \qquad \text{in 'Volts'}$$

where,

B = Magnetic flux density   in 'Weber / m^2'

L = Length of conductor (fluid)

   = diameter of pipe   in 'm'

V = Velocity of conductor (fluid)   in 'm/sec.'

- This induced emf E is collected by the electrode and it is given to the external circuit.
- Since, this induced emf E is assumed to be directly proportional to velocity of flowing fluid, therefore, the emf so induced or produced becomes a measure of flow.

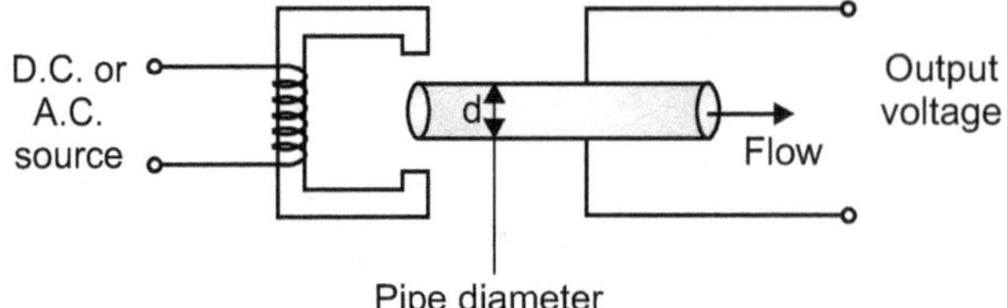

Fig. 3.11: Electromagnetic Flow Meter

3.13.1 Advantages of Electromagnetic Flow Meter

Major Advantages of Electromagnetic Flow Meter:
1. Existing pipes can be converted in to flow meters by adding a pair of electrodes externally and introducing a magnetic field.
2. No obstruction to flow.
3. High accuracy and better reliability.
4. Linearity over wide range of flow measurements.
5. Fast or rapid response to change in flow.
6. No need of any obstruction in fluid flow, hence, no problem of error arising due to pressure drop.
7. It is suitable for laminar as well as turbulent flow.
8. Preferred in flow measurement of slurries, corrosive and greasy liquids, and liquids containing suspended particles or impurities.
9. Bidirectional flow measurement is possible with this meter.
10. Measurement is independent of viscosity, density, pressure and temperature.

3.13.2 Disadvantages of Electromagnetic Flow Meter

Major Drawbacks of Electromagnetic Flow Meter:
1. If the flow rate to be measured is low, then the emf or voltage produced is very small. Therefore, amplifiers are to be used for conversion of induced voltages into usable form.
2. It works only for those fluids, which are electrical conductors.
3. Highly expensive.
4. Trapped gas bubbles cause errors.
5. Some portion of liquids flowing through pipe may be deposited in the form of coating on inner surface of pipe. It will lead to fouling of electrodes, due to which, output signal obtained will be less as compared to actual value. If fouling is considerably more, then there are more chances of errors in measured value of flow.

3.13.3 Applications of Electromagnetic Flow Meter

1. Used for measuring flow of fluid like corrosive acid.
2. Widely used for flow measurement of cement slurries, sewage, paper pulp, detergents, greases and sticky fluids.
3. Useful for any electrically conducting fluid.
4. It is known as "**No Obstruction to Flow**" meter, hence suitable for highly accurate and reliable measurement of flow of slurries, greasy materials and liquid containing suspended matter.

3.14 TRANSIT TIME ULTRASONIC FLOW METER

- Ultrasonic flow meter works on the basis of Acoustic method of flow measurement, which states that, **"Velocity of an acoustic wave and flow velocity can be added algebraically or summed vectorially".**

Principle of Operation:

- *"It is based on the apparent change in the velocity of propagation of ultrasonic wave pulses in a fluid with a change in velocity of fluid flow".*

Construction:

- Ultrasonic transducer may be mounted in a direction parallel to the pipe wall or in an inclined direction, making some angle across the pipe wall.
- Ultrasonic transducer consists of two transmitters and two receivers.
- They are separated by a distance "L" and mounted as shown in Fig. 3.12.
- Transmitter "A" & "B" are piezoelectric crystals or devices transmitting short duration ultrasonic signals through the fluid flowing through pipe at a velocity "V".
- Similar types of piezoelectric crystals or devices are used as receivers. Therefore, they are also labelled or referred as "A" and "B".

Working:

- Transmitter "A" transmits wave pulses of short duration in the direction of receiver "A". This transmission is favoured, because, it occurs in the direction of flow of fluid.
- Transmitter "B" transmits wave pulses of short duration in the direction of receiver "B". This transmission is not favoured, because, it occurs in opposite direction of flow of fluid.
- In short, Velocity of ultrasonic waves is increased or decreased by the fluid velocity depending upon the direction of flowing fluid.
- Velocity of ultrasonic signal transmitted by transmitter "A" and received by receiver "A" will increase, due to the fluid velocity "V" (Same direction).
- Whereas, velocity of ultrasonic signal transmitted by transmitter "B" and received by receiver "B" will reduce due to fluid velocity "V" (Opposite direction).

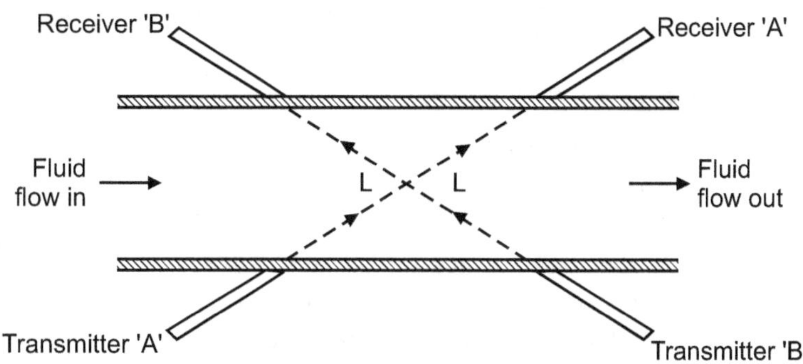

Fig. 3.12: Ultrasonic Flow Meter

- Since, transducers are used as both, transmitters and receivers, the difference in travel time can be determined with same pair of transducers.
- In this device, the detector measures the time taken by ultrasonic wave to cross the pipe along the direction of flow and opposite to the direction of flow.

Let,

V_s = Velocity of sound in fluid	in 'm/s'
V = Velocity of fluid	in 'm/s'
L = Distance between transmitter and receiver	in 'm'
t_1 = Transit time along the fluid flow	in 'seconds'
t_2 = Transit time against the fluid flow	in 'seconds'

$$\textbf{Velocity of flow} = V = \frac{V_s^2\,(t_2 - t_1)}{2 \times L}$$

- **Range of Ultrasonic Meter:** 0 to 180 km/hr on air with an accuracy of ± 0.5%.

3.14.1 Advantages of Transit Time Ultrasonic Flowmeter

1. No need of obstructing fluid flow.
2. No moving parts.
3. It can be used for bi-directional flows as well as rapidly varying/ pulsating flows.
4. High accuracy.
5. Fast response.
6. Linear relationship between velocity of fluid flow and the output obtained.
7. Flow measurement is not affected by fluid viscosity, pressure and temperature variation.
8. Suitable for flow measurements of liquids as well as gases.

3.14.2 Disadvantages of Transit Time Ultrasonic Flowmeter

1. Complex in construction (complex circuit).
2. Relatively high cost.
3. Limited use in industry.
4. Close tolerances (small tolerance zone) are required in manufacturing of mechanical and electronic components, which leads to increase in manufacturing cost.

3.14.3 Applications of Transit Time Ultrasonic Flowmeter

1. To measure the flow rate of any liquid, through which, sound can be transmitted.
2. To measure ocean currents.
3. To measure water flow and gas flow in large conduits (pipes).
4. To measure flow of various industrial fluids, which cannot be measured by other meters.

3.15 DOPPLER TYPE ULTRASONIC FLOW METER

Working Principle:

- Doppler flow meter works on the **principle of doppler frequency shift.**
- It works on the principle, which states that, *"Particles suspended in the fluid impart a frequency shift proportional to the particle velocity, which can be measured with a suitable electronic counter."*

Construction and Working:

- Fig. 3.13 shows construction of Doppler flow meter, in which, one crystal transducer is mounted outside the pipe.
- This transducer emits an ultrasonic wave and the wave is projected at an angle through the pipe wall into the liquid.
- The transducer is basically piezoelectric crystal with a heavy backing to reduce the effect of unwanted rear movement.

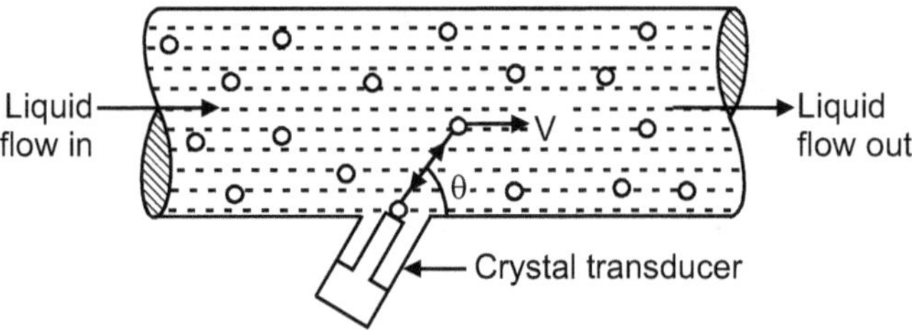

Fig. 3.13: Doppler Type Ultrasonic Flow meter

- Some part of the ultrasonic wave is reflected back by bubbles present in the liquid and liquid particles. This reflected part of ultrasonic wave is returned towards the transducer through the pipe wall.
- Since the reflector (bubbles) is travelling at the fluid velocity, the frequency of the reflected wave is shifted according to doppler principle.
- Velocity of fluid is given by,

$$V = \frac{V_s \cdot \Delta f}{2f_o \cdot \cos(\theta)} = \Delta f \cdot K$$

where, V_s = Velocity of sound in m/sec.

Δf = Difference between transmitted and received frequency

f_o = Frequency of transmission

θ = Angle of transmitter and receiver crystal with respect to axis of pipe in degrees

K = A constant = $\dfrac{V_s}{2 f_o \cdot \cos(\theta)}$

3.15.1 Advantages of Doppler Type Ultrasonic Flow Meter

1. No obstruction to flow.
2. It is not sensitive to variations in viscosity, temperature, density etc.

3. No moving parts, so less or even no wear.
4. Life is more.
5. Excellent dynamic response.

3.15.2 Disadvantages of Doppler Type Ultrasonic Flow Meter

1. High cost.
2. Complex circuit.

3.15.3 Application of Doppler Type Ultrasonic Flow Meter

- It is used mainly for flow measurement of liquids without any pressure loss.

3.15 FACTORS TO BE CONSIDERED WHILE SELECTING A FLOWMETER

1. Measuring range.
2. Chemical compatibility.
3. Accuracy requirements.
4. Pressure requirements.
5. Acceptable pressure drop.
6. Calibration facilities.
7. Desired measurement units (such as volume flow rate, velocity or mass flow rate).
8. Fluid viscosity limitations.
9. Quality of fluid.
10. Properties of the flowing fluid.
11. Cost consideration.
12. Installation.
13. Accessibility and ease of maintenance.

Table 3.1: Names of Flow Measuring Devices Most Suitable for Different Types of Flow

Sr. No.	Type of flow/fluid	Device used
1.	Flow of exceptionally clean fluid only	Rotameter
2.	Flow of fluid containing solid matter	Orifice meter
3.	Water flow in river	Ultrasonic flow meter
4.	Ocean current	Ultrasonic flow meter
5.	Fluid having rapidly fluctuating velocity	Ultrasonic flow meter

Table 3.2: Various Types of Flowmeters with Their Operating Principle and Applications

Sr. No.	Type of Flow Meter	Operating Principle	Application
1.	**Variable area flowmeter (Rotameter)**	Keeping differential pressure constant, change in the area of annular orifice is proportional to flow rate.	For measurement of: 1. Flow of liquid. 2. Flow of gases.
2.	**Variable head flowmeters (Venturimeter, orifice meter)**	A restriction to flow of fluid flowing through pipe is done by introducing venture tube or orifice plate. This restriction creates local pressure drop, which is directly proportional to flow rate.	For measurement of: 1. Flow of fluids in industries. 2. Flow of fluids in commercial applications.

contd. ...

3.	Hot wire anemometer	When fluid flows over a heated surface, heat is transferred from the surface and therefore, its surface temperature reduces. The rate of reduction of temperature is in proportion to the flow rate.	For measurement of: 1. Flow of gases having rapidly fluctuating velocities.
4.	Positive displacement meter	As fluid flows through the meter, it separates the flow of liquid into separate volumetric increments, which are counted and totaled. The sum of increments gives the measurement of the total volume of liquid passed through the meter.	For measurement or metering of: 1. Gasoline and crude oil. 2. Society water.
5.	Ultrasonic flowmeter	The difference in transit times of ultrasonic pulses is linearly proportional to flow velocity.	For measurement of 1. Flow of liquid in industrial applications.

Solved Numerical Type I – Venturimeter

Problem 3.1: *A Venturimeter has an area ratio 9 : 1, the larger diameter being 300 mm. During the flow, the recorded pressure in the large section is 6.5 m and that at the throat is 4.25 m. If the meter coefficient C_d = 0.99, calculate the discharge through the meter.*

Solution: Given data: $\dfrac{A_1}{A_2}$ = 9, d_1 = 300 mm = **0.3 m**, C_d = 0.99

$$h = \frac{P_1}{\rho g} - \frac{P_2}{\rho g} = 6.5 - 4.25 = \textbf{2.25 m}$$

Procedure:

$$A_1 = \frac{\pi}{4} d_1^2 = \frac{\pi}{4} \times (0.3)^2 = \textbf{0.07068 m}^2$$

Given: $\qquad \dfrac{A_1}{A_2} = 9$

$\therefore \qquad\qquad A_2 = \dfrac{A_1}{9} = \dfrac{0.07068}{9} = \textbf{0.007853 m}^2$

Discharge is given by,

$$Q_{act} = \frac{C_d A_1 \cdot A_2 \sqrt{2gh}}{\sqrt{A_1^2 - A_2^2}}$$

$$= \frac{0.99 \times 0.07068 \times 0.007853 \times \sqrt{2 \times 9.81 \times 2.25}}{\sqrt{(0.07068)^2 - (0.007853)^2}}$$

$$= \textbf{0.05198 m}^3\textbf{/sec}$$

Problem 3.2: *A horizontal venturimeter with inlet and throat diameters 30 cm and 15 cm respectively is used to measure the flow of water. The reading of differential manometer connected to inlet and throat is 20 cm of Hg. Determine rate of flow. Take C_d = 0.98.*

Solution: Given data: d_1 = 30 cm = **0.3 m**, d_2 = 15 cm = **0.15 m**,

x = 20 cm = **0.2 m**, C_d = 0.98, S_{Hg} = **13.6** = S_h, S_{water} = **1** = **S**.

Procedure:

$$A_1 = \frac{\pi}{4} d_1^2 = \frac{\pi}{4} \times (0.3)^2 = \textbf{0.07068 m}^2$$

$$A_2 = \frac{\pi}{4} d_2^2 = \frac{\pi}{4} \times (0.15)^2 = \textbf{0.01767 m}^2$$

Also,
$$h = x \cdot \left[\frac{S_h}{S} - 1\right]$$

$$\therefore \quad h = 0.2 \times \left[\frac{13.6}{1} - 1\right] = \mathbf{2.52\ m}$$

Rate of flow or Discharge,
$$Q_{act} = \frac{C_d\, A_1 A_2\, \sqrt{2gh}}{\sqrt{A_1^2 - A_2^2}}$$

$$= \frac{0.98 \times 0.07068 \times 0.01767\, \sqrt{2 \times 9.81 \times 2.52}}{\sqrt{(0.07068)^2 - (0.01767)^2}}$$

$$Q_{act} = \mathbf{0.1257\ m^3/sec}$$

Problem 3.3: *A 300 mm × 150 mm venturimeter is inserted in a vertical pipe carrying water flowing in upward direction. A differential mercury manometer connected to inlet and throat gives reading of 200 mm. Find discharge through pipe. Take C_d = 0.98.*

Solution: Given data: d_1 = 300 mm = **0.3 m**,

d_2 = 150 mm = **0.15 m**,

x = 200 mm = **0.2 m**, C_d = 0.98

Procedure:

Area at inlet of pipe, $\mathbf{A_1} = \frac{\pi}{4}\, d_1^2 = \frac{\pi}{4} \times (0.3)^2 = \mathbf{0.0707\ m^2}$

Area at outlet of pipe, $\mathbf{A_2} = \frac{\pi}{4}\, d_2^2 = \frac{\pi}{4} \times (0.15)^2 = \mathbf{0.0177\ m^2}$

Values of specific gravity of water flowing through the pipe and Mercury (Manometric fluid) are not given in Numerical problem. Let us assume their standard values as,

$$S_{water} = 1$$
$$S_{Hg} = 13.6$$

To calculate difference of pressure head (h), we have,

$$h = x \cdot \left[\frac{S_{Hg}}{S_{water}} - 1\right]$$

$$\therefore \quad h = 0.2 \times \left[\frac{13.6}{1} - 1\right] = \mathbf{2.52\ m}$$

Using venturimeter, discharge can be calculated as,

$$Q = \frac{C_d \cdot A_1 \cdot A_2\, \sqrt{2gh}}{\sqrt{A_1^2 - A_2^2}}$$

$$\therefore \quad Q = \frac{0.98 \times 0.0707 \times 0.0177 \times \sqrt{2 \times 9.81 \times 2.52}}{\sqrt{(0.0707)^2 - (0.0177)^2}}$$

$$\therefore \quad Q = \mathbf{0.12597\ m^3/sec} = 0.12597 \times 10^3 \text{ litres/sec}$$

$$\therefore \quad Q = \mathbf{125.97\ litres\ per\ second\ (\mathit{lps})}$$

Problem 3.4: *An oil of specific gravity 0.8 is flowing through venturimeter having inlet diameter 20 cm and throat diameter 10 cm. The oil-mercury differential manometer shows a reading of 25 cm. Calculate discharge of oil through the horizontal venturimeter. Take C_d = 0.98.*

Solution: Given data: d_1 = 20 cm = **0.2 m**, d_2 = 10 cm = **0.1 m**, x = 25 cm = **0.25 m**, C_d = 0.98, S_{Hg} = **13.6** = $\mathbf{S_h}$, S_{oil} = **0.8** = **S**.

Procedure: $\mathbf{A_1} = \frac{\pi}{4}\, d_1^2 = \frac{\pi}{4} \times (0.2)^2 = \mathbf{0.03141\ m^2}$

$$\mathbf{A_2} = \frac{\pi}{4}\, d_2^2 = \frac{\pi}{4} \times (0.1)^2 = \mathbf{0.007853\ m^2}$$

Also,
$$h = x \cdot \left[\frac{S_h}{S} - 1\right]$$

$$\therefore \quad h = 0.25 \times \left[\frac{13.6}{0.8} - 1\right] = \mathbf{4\ m}$$

$$\text{Discharge, } Q_{act} = \frac{C_d \cdot A_1 A_2 \sqrt{2gh}}{\sqrt{A_1^2 - A_2^2}}$$

$$\therefore \quad Q_{act} = \frac{0.98 \times 0.03141 \times 0.007853 \times \sqrt{2 \times 9.81 \times 4}}{\sqrt{(0.03141)^2 - (0.007853)^2}}$$

$$\therefore \quad Q_{act} = \mathbf{0.07041\ m^3/sec.}$$

Solved Numerical Type II – Orificemeter

Problem 3.5: *An orifice meter with orifice having diameter 15 cm is inserted in a pipe of 30 cm diameter. The pressure difference measured by mercury oil differential manometer is 50 cm of Hg. Find rate of flow of oil having specific gravity 0.9. Take C_d = 0.64.*

Solution: Given data: d_1 = 30 cm = **0.3 m, d_2** = 15 cm = **0.15 m**

$$\mathbf{x} = 50 \text{ cm of Hg} = \mathbf{0.5\ m\ of\ Hg}, \ S_{oil} = \mathbf{0.9} = \mathbf{S}, \ S_{Hg} = \mathbf{13.6} = \mathbf{S_h},$$
$$C_d = 0.64$$

Procedure:
$$\mathbf{A_1} = \frac{\pi}{4}(d_1)^2 = \frac{\pi}{4} \times (0.3)^2 = \mathbf{0.07068\ m^2}$$

$$\mathbf{A_2} = \frac{\pi}{4}(d_2)^2 = \frac{\pi}{4} \times (0.15)^2 = \mathbf{0.01767\ m^2}$$

We know,
$$h = x \cdot \left[\frac{S_h}{S} - 1\right] = 0.5 \times \left[\frac{13.6}{0.9} - 1\right] = \mathbf{7.055\ m}$$

Rate of flow (discharge):

We have,
$$Q_{act} = \frac{C_d A_1 A_2 \sqrt{2gh}}{\sqrt{A_1^2 - A_2^2}}$$

$$\therefore \quad Q_{act} = \frac{0.64 \times 0.07068 \times 0.01767 \times \sqrt{2 \times 9.81 \times 7.055}}{\sqrt{(0.07068)^2 - (0.01767)^2}}$$

$$\therefore \quad Q_{act} = \mathbf{0.1374\ m^3/sec}$$

Problem 3.6: *An orificemeter with orifice diameter 12 cm is inserted in a pipe of 20 cm diameter. The pressure difference between upstream and downstream of the orifice is 9.81 N/cm².The coefficient of discharge for orificemeter is 0.6. Find the discharge of water through the pipe.*

Solution: Given data: d_1 = 20 cm = **0.2 m, d_2** = 12 cm = **0.12 m**,

$$\mathbf{P_1 - P_2} = 9.81 \text{ N/cm}^2 = \mathbf{9.81 \times 10^4\ N/m^2}.$$

Procedure:
$$\mathbf{A_1} = \frac{\pi}{4} \times (d_1)^2 = \frac{\pi}{4} \times (0.2)^2 = \mathbf{0.03141\ m^2}$$

$$\mathbf{A_2} = \frac{\pi}{4} \times (d_2)^2 = \frac{\pi}{4} \times (0.12)^2 = \mathbf{0.0113\ m^2}$$

Pressure difference between upstream and downstream of orifice meter can be written as,
$$P_1 - P_2 = \rho g h$$
where,
$$h = \text{Pressure head difference in 'm'}$$
$$\therefore \quad 9.81 \times 10^4 = 1000 \times 9.81 \times h$$
$$\therefore \quad h = \mathbf{10\ m}$$

Discharge or rate of flow:

$$Q_{act} = \frac{C_d \cdot A_1 A_2 \sqrt{2gh}}{\sqrt{A_1^2 - A_2^2}}$$

$$\therefore \quad Q_{act} = \frac{0.6 \times 0.03141 \times 0.0113 \times \sqrt{2 \times 9.81 \times 10}}{\sqrt{(0.03141)^2 - (0.0113)^2}}$$

$$\therefore \quad Q_{act} = \mathbf{0.10178\ m^3/sec}$$
$$\therefore \quad Q_{act} = \mathbf{101.78\ lps} \qquad [\because \ 1 \text{ litre} = 1 \times 10^{-3}\ m^3]$$

Solved Numerical Type III – Pitot Tube

Problem 3.7: *A Pitot tube is inserted into a water stream having velocity 2.3 m/sec. It has gauge difference of 30 cm on the water-mercury manometer. Find the coefficient of velocity.*

Solution: Given data: V = 2.3 m/s, **x** = 30 cm = **0.03 m**, S_{Hg} = **13.6** = S_h and

S_{water} = **1** = **S**.

Procedure:

We have,
$$h = x \cdot \left[\frac{S_h}{S} - 1\right] = 0.03 \times \left[\frac{13.6}{1} - 1\right] = \textbf{0.378 m}$$

We know,
$$V_{act} = C_v \sqrt{2gh}$$

∴
$$C_v = \frac{V_{act}}{\sqrt{2gh}} = \frac{2.3}{\sqrt{2 \times 9.81 \times 0.378}} = \textbf{0.8445}$$

Problem 3.8: *If the Pitot tube shows 15.5 cm of water, determine the velocity of water. If area of flow is 5.88 cm^2, determine the discharge of water in lit/sec. Take C_v = 0.98.*

Solution: Given data:
$$h = 15.5 \text{ cm} = \textbf{0.155 m}$$
$$A = 5.88 \text{ cm}^2 = \textbf{5.88} \times \textbf{10}^{-4} \textbf{ m}^2$$

Procedure:

1. Velocity of water:

Velocity of water = V_{act} = $C_v \sqrt{2gh}$
$$= 0.98 \times \sqrt{2 \times 9.81 \times 0.155} = \textbf{1.709 m/s}$$

2. Discharge: Q = A × V_{act}
$$= 5.88 \times 10^{-4} \times 1.709 = 1 \times 10^{-3} \text{ m}^3/\text{sec} = \textbf{1 } \textit{lps}$$

Problem 3.9: *A pitot-static tube is used to measure the velocity of water in a pipe. The stagnation pressure head is 6 m and static pressure head is 5 m. Calculate the velocity of flow, assuming coefficient of tube equal to 0.98.*

Solution: Given data: C_v = 0.98
$$h = (\text{Stagnation pressure head}) - (\text{Static pressure head})$$

∴
$$h = 6 - 5 = \textbf{1 m}$$

Procedure:

Velocity of flow, V = $C_v \sqrt{2gh}$
$$= 0.98 \times \sqrt{2 \times 9.81 \times 1} = \textbf{4.34 m/s}$$

Problem 3.10: *Find the velocity of an oil flowing through a pipe, when the difference of mercury level in a differential U-tube manometer connected to two tapings of the pitot tube is 100 mm. Take S.G. of oil as 0.8 and coefficient of pitot tube as 0.98.*

Solution: Given data: x = 100 mm = **0.1 m**, S_{oil} = **0.8** = **S** and S_{Hg} = **13.6** = S_h, C_v = 0.98.

Procedure: We know, **h** = $x \cdot \left[\dfrac{S_h}{S} - 1\right] = 0.1 \times \left[\dfrac{13.6}{0.8} - 1\right] = \textbf{1.6 m}$

Velocity of oil, **V** = $C_v \sqrt{2gh} = 0.98 \times \sqrt{2 \times 9.81 \times 1.6} = \textbf{5.49 m/s}$

Important Points

- **Flow measurement** is defined as, "the process of measuring amount of fluid flowing through a given point in a given (definite) time period."
- **Rate of flow** or **Discharge** is defined as, "the quantity of a liquid, flowing per second through a section of a pipe."
- **Flow meter** is capable of measuring quantity of fluid flowing in a circular pipe.

- **Rate meters (Inferential):** The term **"Inferential"** is applied to those flowmeters, which do not measure the flow directly. Such type of flow meters measure other quantities, such as, pressure, temperature etc., which are related to flow. Therefore, they are called as "**Inferential type flow meters**".

- **Working principle of variable head meters:** "If a restriction element (obstruction) such as orifice plate or venturi-tube or elbow in the pipe of flowing fluid is introduced, then it produces a pressure difference across the restriction element used."

- **Different types of variable head type flow meters are,**
    - (i)   Venture tube or Venturimeter,
    - (ii) Orifice plate or orifice meter,
    - (iii) Pitot tube.

- **Venturimeter** is a device used to measure rate of discharge in a pipeline. It is fixed permanently at different sections of pipeline to know the discharges at those sections.

- **Orifice meter** works on same principle as that of venturimeter i.e. the pressure head (energy) is converted into kinetic energy in an accelerated flow.

- **Orifice** is used in orifice meter, which is used to measure discharge of fluid flowing through the pipe.

- **A U-tube differential manometer** is connected between two sections, one on upstream side and other at vena contracta on the downstream side of orifice plate.

- **Pitot tube** is used to measure the velocity of fluid flowing through a pipe at any point.

- **Working principle of pitot tube:** "If velocity of flow at a point becomes zero, there is an increase in pressure energy". This point is known as stagnation point.

- **Dall tube** is a shortened version of a Venturi meter. Dall Flow Tube consists of two sections with relatively large cone angles. It has a short straight inlet section which is followed by an abrupt decrease in the inside diameter of the tube. A narrow annular slit separates the short inlet and divergent outlet. The throat is formed by a circumstantial slit located between inlet and outlet cones.

- **Principle of working of rotameter:** "Pressure difference across the orifice is directly proportional to square of its flow area and square of flow rate."

- **Working principle of hot wire anemometer:** "When fluid flows over a heated surface, the surface temperature reduces due to transfer of heat from the heated surface to the fluid. This rate of reduction in temperature is a measure of flow rate."

- **Oscillating piston type flow meter** consists of a slotted cylinder which oscillates about a dividing bridge. This separates the inlet port from the outlet port.

- **Rotating vane flow meter** consists of an eccentric drum carrying a number of spring-loaded radial vanes. Fluid pressure rotates the drum and vanes get pushed outward by springs to form separated sealed chambers. During drum rotation, a fixed quantity of fluid is trapped inside each chamber and is successively delivered to outlet.

- **Working principle of doppler flow meter:** "Particles suspended in the fluid impart a frequency shift proportional to the particle velocity, which can be measured with a suitable electronic counter."

- **Working principle of transit time ultrasonic flow meter:** Ultrasonic flow meter works on the basis of Acoustic method of flow measurement, which states that, "velocity of an acoustic wave and flow velocity can be added algebraically or summed vectorially". Its working is based on the apparent change in the velocity of propagation of ultrasonic wave pulses in a fluid with a change in velocity of fluid flow.

Theory Questions For Practice

1. Enlist different flow transducers. Explain the working of rotameter.
2. Draw neat sketch of rotameter and explain its working.
3. Explain the working of rotameter with the help of neat diagram.
4. Draw labelled sketch of hot wire anemometer and explain its working.
5. Explain construction and working of hot wire anemometer.
6. Explain the working principle of hot wire anemometer in constant current and constant temperature mode, with neat sketch.
7. What is ultrasonic flow measurement? Explain its working principle with necessary figure.

8. Draw neat sketch of ultrasonic flow meter and explain how flow is measured by using it.

9. Explain the working of ultrasonic flow meter with a neat sketch.

10. How flow is measured by hot wire anemometer?

11. List suitable applications of flow measurement devices, (i) Rotameter, (ii) Hot wire anemometer, (iii) Ultrasonic flow meter.

12. What is orifice plate? Explain its working.

13. Differentiate between orifice meter and venturimeter.

14. Explain with a neat sketch, working of variable area flow meter.

15. Explain working of velocity flow meter with a neat sketch.

Numerical Problems For Practice

Numerical Type I – Venturimeter

1. A venturimeter is used to measure the rate of flow of a liquid of S.G. 0.8. Inlet diameter of venturimeter is 80 mm and throat diameter is 50 mm. The differential manometer reads 20 cm of Hg. Calculate the discharge, if $C_d = 0.98$. (**Ans.** 5.25 lit/s)

2. A venturimeter 100 mm × 50 mm size is used to measure the flow of liquid of specific gravity 0.8. If the mercury differential manometer head is 200 mm, find the discharge through the venturimeter. (**Ans.** 0.01574 m^3/sec)

Numerical Type II – Orifice meter

3. An orifice meter with orifice diameter 15 cm is inserted in a pipe of 30 cm diameter. The pressure difference measured by mercury-oil differential manometer is 50 cm of Hg. Find rate of flow, if specific gravity is 0.9. Take $C_d = 0.64$. (**Ans.** 0.1293 m^3/sec)

Numerical Type III – Pitot Tube

4. Find the velocity of the liquid flowing, through a pipe, fitted with a pitot tube. The reading of the pitot tube is 20 cm and the value of C_V can be taken as 0.98. (**Ans.** 1.94 m/s)

■■■

Chapter 4

MISCELLANEOUS MEASUREMENT

4.1 DYNAMOMETER

- Dynamometer is a device used to measure the torque being exerted along a rotating shaft and hence, power transmitted by shaft.

4.2 CLASSIFICATION OF DYNAMOMETER

- Dynamometers are classified as,
    1. **Absorption type dynamometer:**
        o Here, the energy is converted into heat by friction at the time of measuring.
        o **Example:** Eddy current dynamometer.
        o **Applications:** Engine, turbine and electric motor.
    2. **Transmission dynamometer:**
        o Here, the power being measured is not absorbed or dissipated to surrounding. But, it is transmitted either to or from dynamometer.
        o Therefore, after measurement, power is available in a useful mechanical or electrical form.
        o **Examples:** Strain gauge transmission dynamometer, Belt dynamometer.
    3. **Driving dynamometer:**
        o These instruments measure power and also supply energy to operate the tested devices.
        o They are coupled to either power absorbing or power developing devices, as they have to operate either as a motor or generator.
        o **Example:** Electric dynamometer.

4.3 EDDY CURRENT DYNAMOMETER (ABSORPTION TYPE DYNAMOMETER)

- It is an electrical absorption type dynamometer.

Principle of Working:

- Eddy Current Dynamometer works on the principle that, *"when a conductor moves through a magnetic field, it induces eddy currents (called as local currents). These eddy currents flow in a short circular path around the conductor and are dissipated in the form of heat."*

Construction:

- Eddy Current Dynamometer consists of non-magnetic steel rotor mounted on the shaft, whose power is to be measured.
- It rotates inside a smooth cast iron stator. Clearance between stator and rotor is very small.
- Stator carries an exciting coil, which is energized by d.c. supply from an external source.
- The stator is mounted, such that, it permits free swing about its axis.
- Stator is provided with a torque arm, to which, a spring balance is attached.

Working:

- Stator winding is excited by a d.c. supply as shown in Fig. 4.1.
- When the solid rotor moves in the magnetic field produced by stator winding, an emf is produced in it, thereby inducing eddy currents.
- These eddy currents oppose the rotation of rotor.
- The moment of resistance is measured by torque arm or brake arm and then torque and hence, shaft power can be calculated.

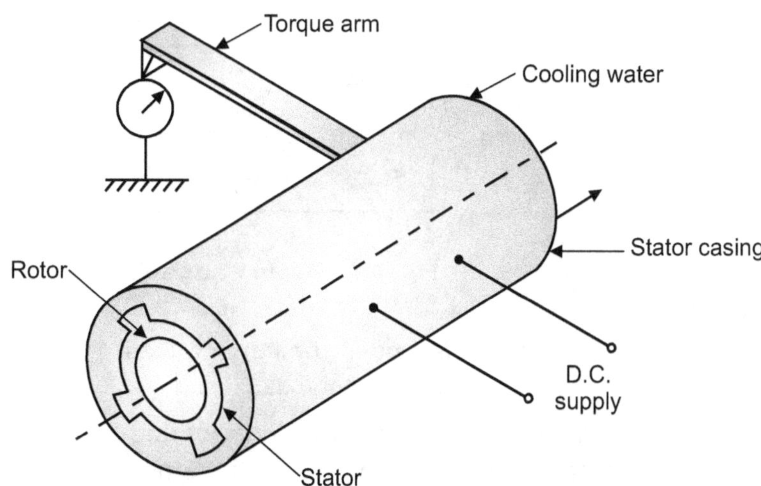

Fig. 4.1: Eddy Current Dynamometer

- The mechanical power supplied to dynamometer is converted in the form of heat.
- This heat is dissipated or carried away partly by air circulation and partly by water. Therefore, cold water is circulated through the air gap between the stator and rotor.

4.3.1 Advantages of Eddy Current Dynamometer

1. Good control at low rotating speeds.
2. Suitable for high speed range.
3. Ability to measure large power output at all speeds. Therefore, it is preferred in testing of automobile and air craft engines.
4. It is compact as compared to other dynamometers of same capacity.
5. The torque developed is smooth and continuous under all operating conditions.

4.3.2 Disadvantages of Eddy Current Dynamometer

1. It cannot produce torque at zero speed.
2. Comparatively smaller in size, which reduces its strength and reliability.

4.3.3 Applications of Eddy Current Dynamometer

1. Measurement of extremely large power in the range of 200 kW-225 kW.
2. If the shafts are rotating at high speeds of 5000 to 6000 r.p.m., it is most suitable instrument for power measurement.

4.4 STRAIN GAUGE TRANSMISSION DYNAMOMETER (TRANSMISSION DYNAMOMETER)

- Fig. 4.2 shows a strain gauge transmission dynamometer configured with a strain gauge Wheatstone bridge circuit. It is widely used for torque measurement. Therefore, it is also called as **strain gauge torsion dynamometer**.

Construction:

- Four bonded-wire strain gauges are mounted at an angle of 45° with respect to the axis of rotation.
- They are placed in **pairs, but diametrically opposite.**
- The gauges are accurately placed, such that, the system is temperature compensated and insensitive to bending, thrust or pull effects. Any change in the gauge circuit then results only from torsional deflection.

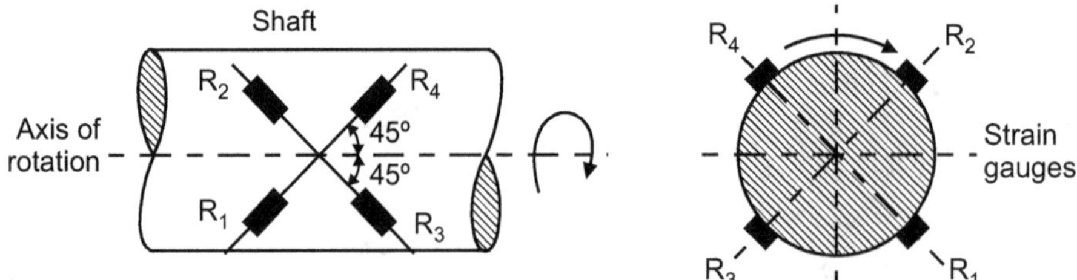

Fig. 4.2: Strain Gauge Transmission Dynamometer

Working:

- When the shaft is under torsion, gauges 1 and 4 will elongate as a result of the tensile component on one diagonal axis.

- Whereas, gauges 2 and 3 will contract due to compressive component on the other diagonal axis.

- These principal strains (tensile and compressive) are measured and hence, torque on shaft is calculated.

- Further, these strain gauges are connected to Wheatstone's bridge circuit. Galvanometer shows the output of Wheatstone's bridge, which is proportional to torsion and hence, to applied torque on shaft.

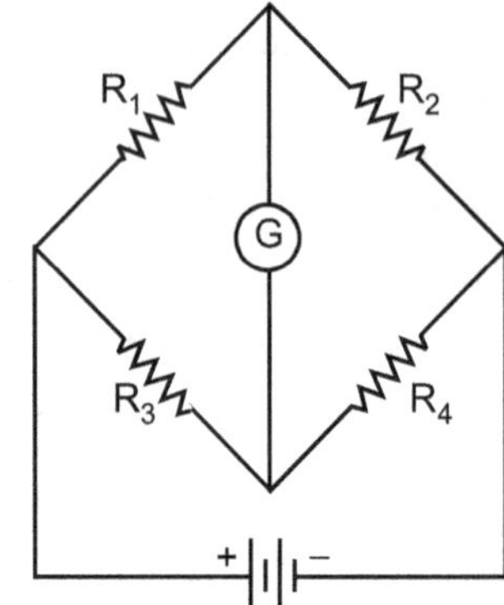

Fig. 4.3: Wheatstone's Bridge Circuit for Strain Gauge Transmission Dynamometer

4.4.1 Advantages of Strain Gauge Torsion Dynamometer

1. Fully temperature compensated.
2. Provides automatic compensation for bending and axial loads.
3. Gives the maximum sensitivity for a given torque.
4. Gives an instantaneous result.

4.4.2 Limitations of Strain Gauge Torsion Dynamometer

1. Expensive.
2. Initial setting needs skilled operator.
3. Initial setting is time consuming.

4.5 MECHANICAL OR LATHE TOOL DYNAMOMETER

Use: The mechanical tool dynamometer measures deflection on the tool holder by the use of sensitive dial indicator.

Construction and Working:

- The mechanical type tool dynamometer consists of sensitive dial indicator for direct measurement of forces. Refer Fig. 4.4.

- The dial indicator is calibrated to read the cutting force.

- The deflection produced by it is directly proportional to the cutting force.

- The dial indicators are placed in direction of cutting force to find deflection.

- The cutting force or tangential force will tend to deflect the tool and tool holder downwards, whereas, axial force or thrust force acting along the axis of the work piece will tend to push the tool away from the work piece.

- The deflection of dial indicator caused by the forces, cutting force and thrust force is calibrated in terms of the force.

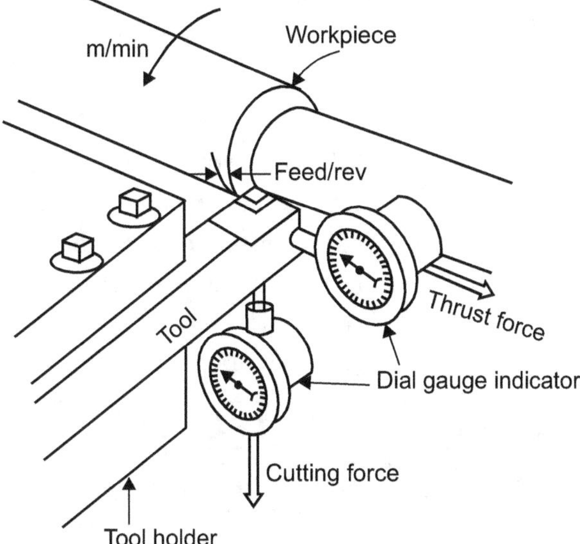

Fig. 4.4: Lathe Tool Dynamometer

4.5.1 Advantages of Mechanical Tool Dynamometer

1. Simple method of tool force measurement.
2. Economical method.

4.5.2 Disadvantages of Mechanical Tool Dynamometer

1. Chances of error.
2. It cannot give digital read out.

4.5.3 Application of Mechanical Tool Dynamometer

1. To find out various cutting forces acting in metal cutting on tool (i.e. axial and tangential forces).

4.6 INTRODUCTION TO SPEED MEASUREMENT

- Speed can be linear or angular according to movement of object.
  1. **Linear movement**:
     - Linear movement means displacement per unit time, which is measured as **Linear speed** (V) in m/sec.
  2. **Rotational movement**:
     - It is measured in,
       - **(i) Number of revolutions per minute** (N) in r.p.m.
       - **(ii) Angular speed** (ω) in rad/sec.

4.7 TACHOMETER

- Tachometers are the most commonly used devices for measurement of speed.
- Tachometer is defined as, **"an instrument used to measure the angular velocity of shaft and express it in number of revolutions made per minute (in r.p.m.)."**
- It may continuously indicate the value of angular speed. Also, it may indicate average angular speed measured within short intervals of time.

4.8 CLASSIFICATION OF TACHOMETERS

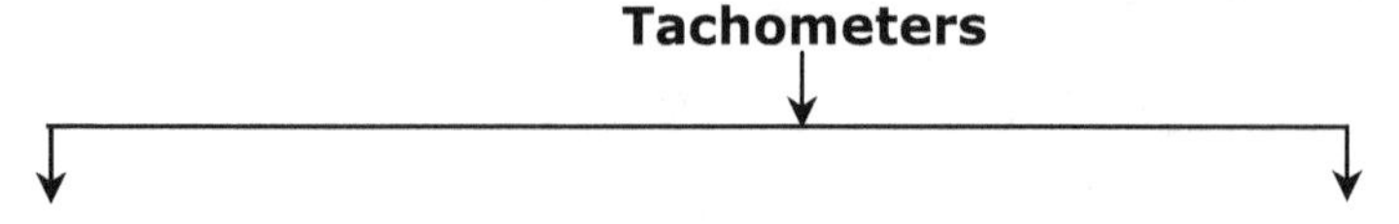

Tachometers

Mechanical Type

(i) Revolution counter and timer
(ii) Slipping clutch tachometer
(iii) Hand speed indicator
(iv) Centrifugal force type tachometer

Electrical Type

(i) Eddy current or Drag type tachometer (Contact type)
(ii) Contactless type
 (a) Inductive type tachometer
 (b) Capacitive type tachometer
 (c) Photoelectric tachometer
 (d) Stroboscope

4.9 MECHANICAL TACHOMETERS

4.9.1 Revolution Counter and Timer

- **Range:** – 2000 to 3000 r.p.m.

Construction:

- Revolution counter and timer consists of a worm and a worm gear (wheel) in mesh.
- The shaft, whose speed is to be measured, is connected to worm.
- When the rotating shaft is engaged with the worm, worm rotates and hence, it drives or rotates the worm gear (wheel) in mesh.
- Arrangement is made in such a way that, one revolution of worm causes the worm gear (wheel) to rotate by one tooth.
- This arrangement employs a timing device, which counts the number of revolutions measured during certain time interval. A pointer is attached to worm gear.

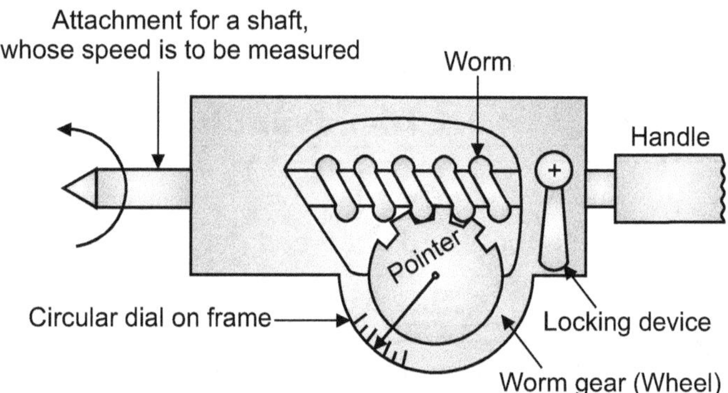

Fig. 4.5: Revolution Counter and Timer

Working:

- When the rotating shaft drives worm, the worm gear (wheel) starts to rotate.
- Rotation of worm gear (wheel) actuates the pointer and makes it to move over calibrated dial (scale).
- Movement of pointer is a measure of shaft speed.
- This movement of pointer can be directly calibrated to give number of revolutions made by the shaft within a specified time. It is done by using a separate timer and stopwatch. Thus, the revolution counter and timer gives the total number of revolutions over a given time, i.e. average rotational speed rather than an instantaneous rotational speed.

4.9.1.1 Advantages of Revolution Counter and Timer

1. Simple to operate.
2. Simple in construction.
3. Mostly used to measure speed of shaft, pulley, gear etc.

4.9.1.2 Disadvantages of Revolution Counter and Timer

1. It is limited to low speed measurements.
2. More chances of error.
3. It gives average speed, instead of instantaneous speed.

4.9.1.3 Application of Revolution Counter and Timer

1. Suitable for speed measurement at definite time intervals, such as, in case of low speed engines.

4.9.2 Slipping Clutch Tachometer

- Slipping clutch tachometer consists of a driving shaft and an indicator shaft, connected by means of slipping clutch.
- The rotating shaft, whose speed is to be measured, drives an indicator shaft.
- The pointer attached to the indicator shaft moves over a calibrated scale and reading obtained is a measure of shaft speed, against the torque of spiral spring.

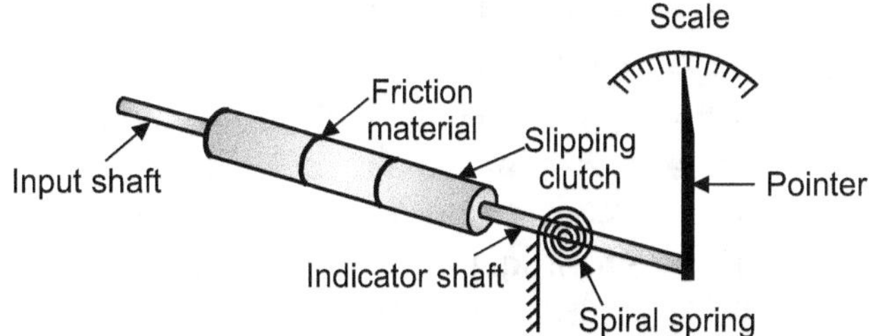

Fig. 4.6: Slipping Clutch Tachometer

4.9.2.1 Advantages of Slipping Clutch Tachometer

1. Simple to operate
2. Simple in construction.

4.9.2.2 Disadvantages of Slipping Clutch Tachometer

1. Limited to low speed measurement.
2. More chance of error.
3. It gives average speed, instead of instantaneous speed.

4.9.2.3 Applications of Slipping Clutch Tachometer

1. Used to measure shaft speed.
2. Preferred in applications, where average speed of shaft is to be measured.

4.9.3 Hand Speed Indicator

- **Range:** 20000 to 30000 r.p.m.
- **Accuracy:** ± 1% of full scale.

Construction:

- Hand speed indicator consists of a spindle having conical tip. Therefore, spindle is also called as **conical spindle.**
- Due to conical tip, spindle can be connected or attached to the shaft, whose angular speed is to be measured.
- **Hand speed indicator** has an integral (inbuilt) stopwatch and counter with automatic disconnect facility (system).

Working:

- When the spindle is brought in contact with the shaft under measurement, hand speed indicator operates, but the counter and hence, the stopwatch do not start functioning, until 'START' or 'RESET' button is pressed.

- As soon as, the button **'START'** or **'RESET'** is pressed, the revolution counter gets engaged with automatic clutch. Now, revolution counter gets actuated and starts functioning.

- The instrument shows the angular speed over short interval of time.

- After measurement, the revolution counter gets automatically disconnected or disengaged. The disengagement takes a fixed time. It is very short duration (approximately 3 to 6 seconds) after performing process of measurement.

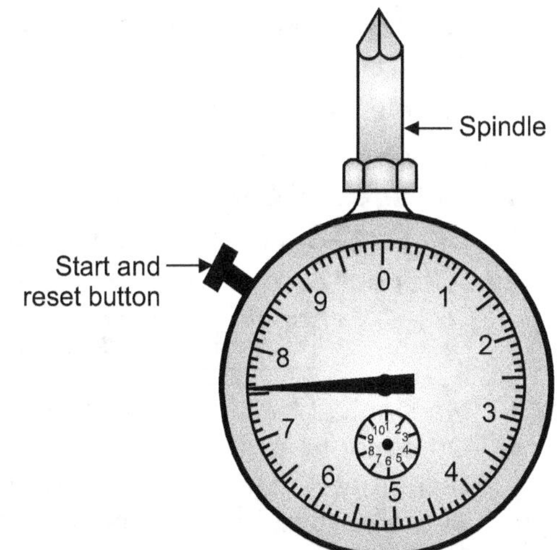

Fig. 4.7: Hand Speed Indicator

4.9.4 Centrifugal Force Tachometer

- **Range:** Up to 40,000 r.p.m.
- **Accuracy:** ± 1 %.
- Centrifugal tachometer is the only mechanical tachometer having **ability to detect the variations in the speed.**

Principle of Working:

- When a body revolves about an axis, it experiences a radially outward force. This force is called as centrifugal force.
- This centrifugal force is a function of rotational speed. Therefore, it is used to measure instantaneous speed by means of centrifugal force.

Construction:

- Two fly balls of small weights are arranged about a central spindle.
- Driver shaft, whose speed is to be measured, is connected to this spindle.

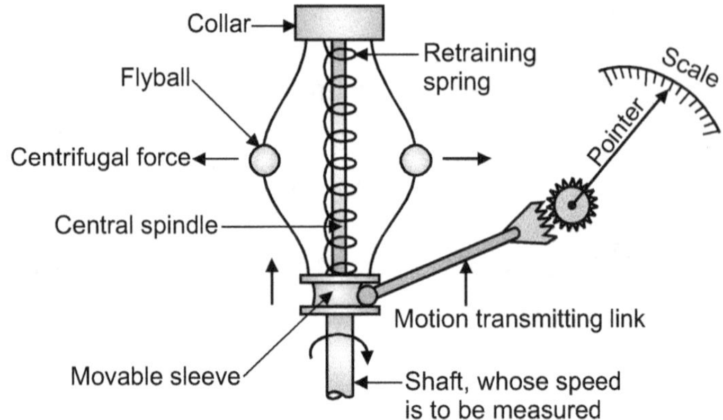

Fig. 4.8: Centrifugal Tachometer

Working:

- As soon as driver shaft is connected to this central spindle, central spindle starts rotating.
- Due to rotation of central spindle, a centrifugal force is developed, which causes the fly balls to move outwards in radial direction, i.e. away from the spindle.
- When fly balls move away from central spindle, spring wounded on the spindle is compressed.
- Value of this compression or decrease in length of spring becomes a function of rotational speed and accordingly, the pointer is moved on a calibrated scale, which is a measure of shaft speed.
- As soon as, driver shaft is disconnected from centrifugal tachometer, the spring is restored to its original length. Fly balls move towards the central spindle and occupy their original position.

Advantages of Centrifugal Tachometer:

1. It has accuracy of ± 1%.
2. It is used to measure speed over wide range.
3. Constant speed is shown.
4. Simple to operate.
5. Easy in maintenance.

Disadvantages of Centrifugal Tachometer:

1. High speed may damage the instrument.
2. It cannot be placed horizontally. It needs to be placed in vertical direction only.

4.9.5 Disadvantages of Mechanical Tachometers

- Mechanical tachometers impose load on the shafts, to which, they are connected.
- On account of inertia of mechanical parts, the mechanical tachometers (except centrifugal tachometer) are not able to detect the variations in the speed. They indicate only the average speed over a specific period of time.
- Mechanical tachometer normally uses an arrangement, where a counter is used for counting the number of revolutions and a stop-watch is used for measurement of time. The speed is calculated on the basis of these measurements. This involves large error, especially, when stopwatch is operated manually.

4.10 ELECTRICAL TACHOMETER

- Electrical tachometer converts the rotational speed into electrical signal, which can be measured and calibrated to give rotational speed directly.
- In short, it is a transducer, which produces an electrical signal in proportion to the rotational speed.
- Depending upon the type of transduction, electrical tachometers are classified into various types. Some of them are explained below.

4.10.1 Eddy Current or Drag Cup Tachometer

Range: Up to 12000 r.p.m.

Accuracy: ± 3 % of full scale.

Construction:

- It is a contact type of electrical tachometer, i.e. physical contact is made with the shaft, whose speed is to be measured.
- In this type of tachometer, the transducer produces an analog signal in the form of continuous drag due to eddy current, induced in the cup.
- The current, which is indicated by an analog indicator on a scale, is proportional to the speed.
- The machine shaft, whose speed is to be measured, is engaged with the tacho-meter shaft. This tacho-meter shaft is connected to a permanent magnet.
- The magnet is surrounded by a cup made up of non-magnetic, but conductive material, such as, aluminum.

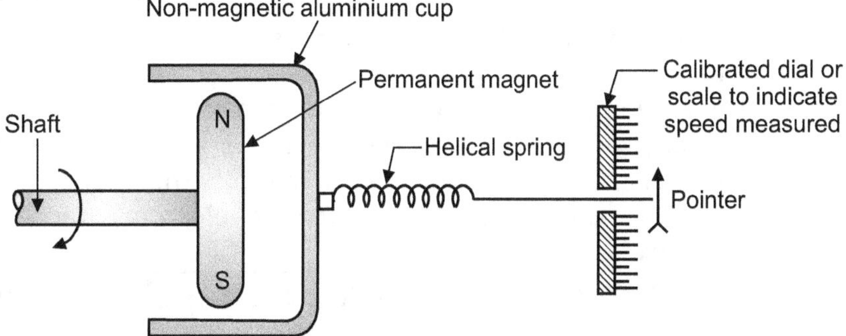

Fig. 4.9: Eddy Current or Drag Type Tachometer

- The other end of non-magnetic aluminum cup is connected to the pointer through a helical spring.

Working:

- Due to rotation of machine shaft, the connected tachometer shaft and hence, the magnet starts rotating. This rotation of magnet induces eddy currents and hence, emf in the non-magnetic aluminum cup.
- The induced eddy currents produce a twisting moment or torque in the cup, which acts against the torque of helical spring.
- Due to the induced torque, the cup turns in the direction of rotating magnet, till the torque induced (developed) becomes equal to torque on helical spring.
- The pointer attached to the cup moves over a calibrated scale (dial), indicating the angular speed of shaft.

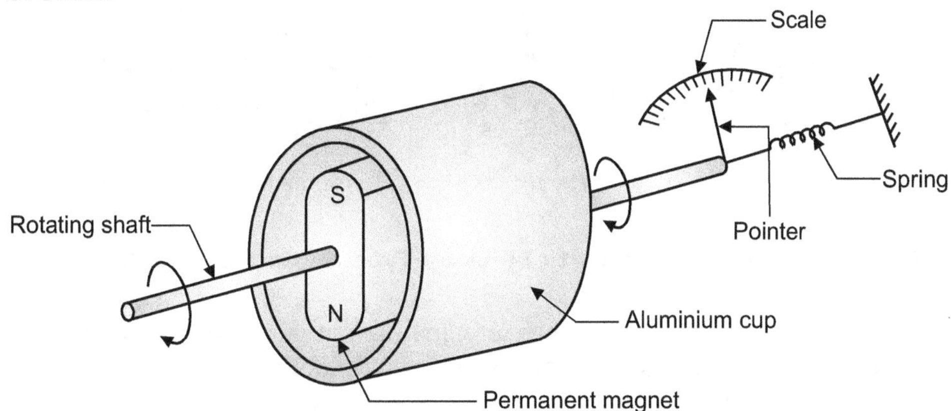

Fig. 4.10: Illustrative View of Eddy Current or Drag Type Tachometer

Note: Eddy current is a type of current generated in a conductor (non-magnetic aluminum cup), which lies in the vicinity of magnetic field.

4.10.1.1 Advantages of Eddy Current or Drag Type Tachometer

1. Low cost.
2. Rigid in construction.
3. Very low maintenance.

4.10.1.2 Disadvantages of Eddy Current or Drag Type Tachometer

1. At a very high speed, it gives non-linear characteristics for output voltage and speed of shaft.
2. To calibrate this type of tachometer, it is essential that, input voltage must be maintained absolutely constant.

4.10.1.3 Applications of Eddy Current or Drag Type Tachometer

1. Automobile speedometer.
2. Aircraft engines.
3. Industrial applications.
4. For measuring speed of locomotives.

4.11 CONTACTLESS ELECTRICAL TACHOMETER

- In this type of tachometer, there is no direct contact with the shaft, whose speed is to be measured.
- The tachometer produces pulse from a rotating shaft.
- The energy produced by this device is not sufficient to actuate the indicator mechanism. Therefore, amplifiers are used to magnify the energy produced.
- The different types of contactless electrical tachometers are explained in following subarticles.

4.11.1 Inductive Pick-Up Type Contactless Tachometer
(Variable Reluctance / Magnetic Pick-Up Type Pulse Counter)

Range:
- Upto 72,000 r.p.m.

Construction:
- Inductive pick up type tachometer consists of a small permanent magnet with a coil wound on it. This permanent magnet is also called as **magnetic pick up.**
- This small permanent magnet is placed near a metallic toothed rotor (Ferromagnetic toothed rotor). This metallic toothed rotor is mounted on shaft, whose speed is to be measured.

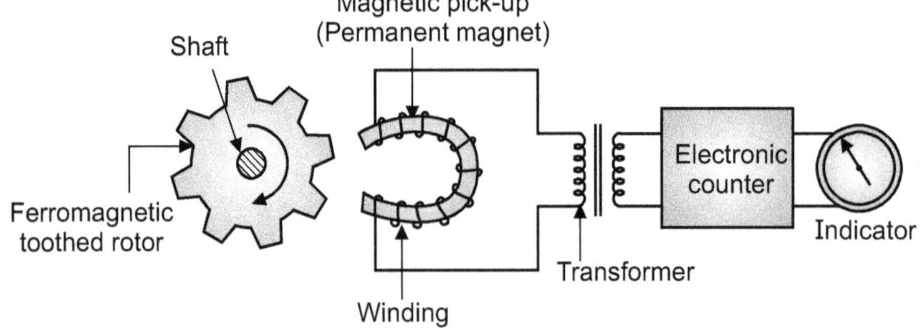

Fig. 4.11: Inductive (Magnetic) Pick up Type Contactless Tachometer

Working:
- When the rotor rotates, the reluctance of air gap between the magnet and toothed rotor changes giving rise to an induced e.m.f. in the pickup coil.
- This output is in the form of pulse.
- The frequency of pulses of induced e.m.f. depends upon the number of teeth on the wheel and its speed of rotation.

- As the numbers of teeth are known and are constant, the speed of rotation can be determined by measuring frequency of pulses with the help of an electronic counter.

- Speed is given by, $\quad N = \dfrac{f}{T} \quad$ in Revolutions per second (r.p.s.)

  Where, $\qquad T =$ Number of teeth on rotor, and

  $\qquad f =$ Number of pulses per second.

- Alternatively, speed can be measured as,

  $$N = \dfrac{f}{T} \times 60 \qquad \text{in Revolutions per minute (r.p.m.)}$$

4.11.1.1 Advantages of Inductive (Magnetic) Pick-up Type Contactless Tachometer

1. Simple in construction.
2. Requires very less maintenance.
3. Easy to calibrate.
4. Very high accuracy.
5. Useful for very low as well as very high-speed measurement.
6. No direct-physical contact is required with shaft, whose speed is to be measured.

4.11.1.2 Disadvantages of Inductive (Magnetic) Pick-up Type Contactless Tachometer

1. Imperfect installation creates problem.
2. Moderate cost.

4.11.1.3 Applications of Inductive (Magnetic) Pick-up Type Contactless Tachometer

1. It is used to measure speed up to 72000 r.p.m.
2. It is used to measure speed of motors, running at very high revolutions per minute.

4.11.2 Capacitive Pick-Up Type Contactless Tachometer

Working Principle:
- Capacitance of a parallel plate capacitor is given by,

$$C = \dfrac{KA}{d}$$

where, $\qquad A =$ Area of each plate or overlapping area of plates $\quad$ in 'm²'

$\qquad d =$ Distance between two plates $\qquad$ in 'm'

$\qquad K =$ Dielectric constant of medium existing between parallel plates

- Thus, we observe that, value of capacitance 'C' will change, if there is change in,
  (i) Overlapping area of plates or
  (ii) Distance between two parallel plates or
  (iii) Dielectric medium and its Dielectric constant

Construction:
- **Capacitive Pick-Up Type Contactless Tachometer** consists of a vane attached to one end of rotating shaft, whose speed is to be measured.

- Two capacitor plates having same cross-sectional area are placed parallel to each other.

- These plates are distantly fixed in such a way that, a small gap exists in between these parallel plates, so that, vane mounted on shaft can pass through this gap.

- These two parallel plates form a parallel plate capacitor.

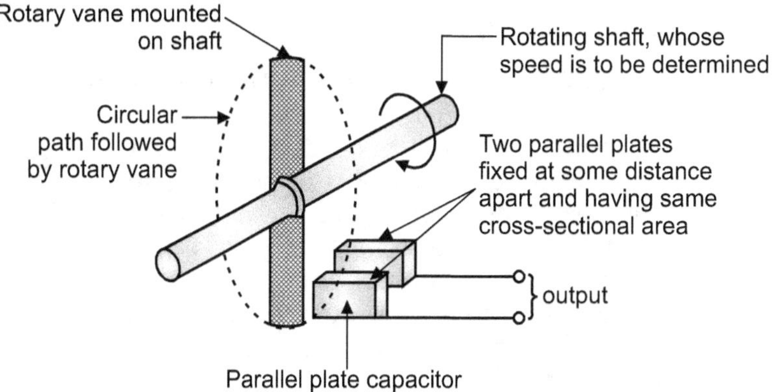

Fig. 4.12: Capacitive Pick-Up Type Contactless Tachometer

Working:

- Both plates have same cross-sectional area. They are placed parallel to each other at a fixed distance. Therefore, overlapping area of plates "A" and distance between two plates "d" will remain constant during the working/operation of this tachometer.
- When shaft rotates, the vane also rotates and hence passes through the air gap existing in between the parallel plates. At this instant, dielectric medium changes from air to vane material. Therefore, value of capacitance changes, due to change in dielectric medium, i.e. dielectric constant (K).
- This change in capacitance can be calibrated in terms of angular speed of shaft on readout device.

4.11.2.1 Advantages of Capacitive Pick-Up Type Contactless Tachometer

1. It is contactless method.
2. High sensitivity.
3. Very small loading effect.

4.11.2.2 Disadvantages of Capacitive Pick-Up Type Contactless Tachometer

1. It has complicated electronic circuit.
2. It is used to measure only small speeds.

4.11.3 Photoelectric Tachometer

- **Range:** Up to 10000 r.p.m.

Construction:

- **Photoelectric Tachometer** uses an **opaque disc** having **equidistant holes** on its **periphery**.
- This disc is mounted on a rotating shaft, whose speed is to be measured.
- A light source is placed on one side of the disc and a light sensor is placed on the other side. The light sensor may be photosensitive device or photo tube.
- The light sensor must be in line with the light source.

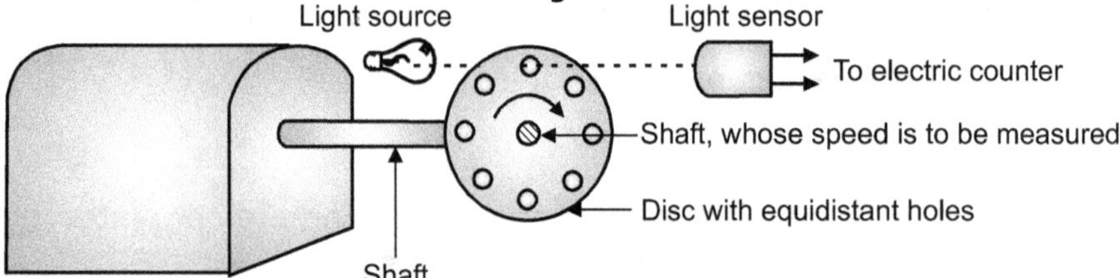

Fig. 4.13: Photoelectric Tachometer

Working:

- On rotation of disc, holes and opaque portions of the disc come alternatively in between light source and light sensor (Receiver).
- Thus, whenever a hole comes in line with light source and sensor, a pulse is generated. Refer Fig. 4.14.
- The frequency of these pulses depends upon (i) Number of holes on discs and (ii) Speed of rotation of disc.
- Here, Speed, $N = \dfrac{f}{H}$

  Where, f = frequency and H = Number of holes.
- When the opaque portion of disc is in between the light source and light sensor, no light falls on the sensor and therefore, no output results are obtained.
- Since the number of holes are fixed, therefore, the number of pulses are counted or measured by means of an electric counter, which is a measure of speed of shaft.
- It is calibrated directly to give speed of shaft in r.p.m.

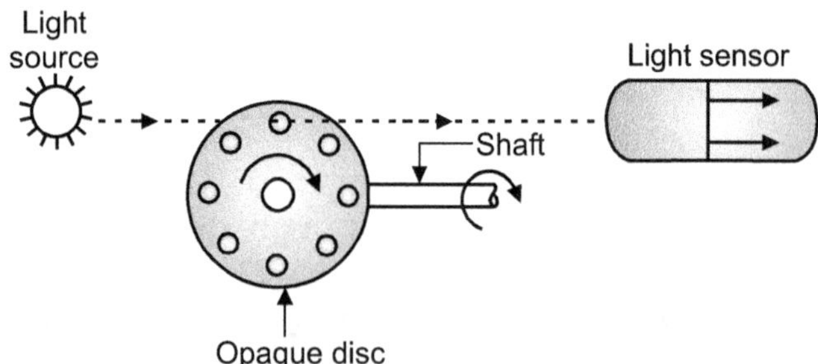

Fig. 4.14: Simple Arrangement of Photoelectric Tachometer

4.11.3.1 Advantages of Photoelectric Tachometer

1. Simple in construction.
2. Output is in digital form, so accuracy is more.
3. Easy to operate and understand.
4. Used to measure very high speed.
5. Less maintenance.
6. It is contactless method.

4.11.3.2 Disadvantages of Photoelectric Tachometer

1. As the life of light source is short, it needs to be replaced periodically.
2. Less accurate.

4.11.3.3 Applications of Photoelectric Tachometer

1. Photoelectric tachometer is used to measure the speed upto 10,000 r.p.m.
2. For speed measurement of motors.

4.11.4 Stroboscope

Principle of Working:

- *"If a strong light is caused to flash on the moving object, the object will appear stationary, if the flashing frequency is equal to speed of moving object."*
- In other words, working principle of stroboscope is **"To synchronize a flashing light with rotation of shaft, making it appears as standstill (stationary)."**

Condition:

- Frequency of flashing light should be equal to angular speed of shaft.
- Therefore, we need a source of light, having provision of changing flashing frequency of light emitted.
- Stroboscope or light source emits light radiations on the circular disc.

Construction:

- Stroboscope is a simple, portable, manually operated device used for measurement of periodic or rotary motion.
- Basically, stroboscope is a source of variable frequency flashing brilliant light.
- The flashing frequency can be set manually by the operator.
- Thus, in short, stroboscope consists of a source of flashing light, whose frequency can be varied or controlled. This source is called as **Strobotron**.
- A distinct or reference mark is made (put) on the shaft (rotating body), whose speed is to be measured.

Working:

- The stroboscope is made to strike light on the mark. Refer Fig. 4.15.
- Circular (Rotary) knob is rotated manually and flashing frequency of light is adjusted, until mark appears stationary.
- When this condition is achieved, it means that, flashing frequency of light is equal to angular speed of shaft.
- Rotation of circular knob can be read from the scale of stroboscope, which is directly calibrated to give angular speed of shaft in r.p.m.

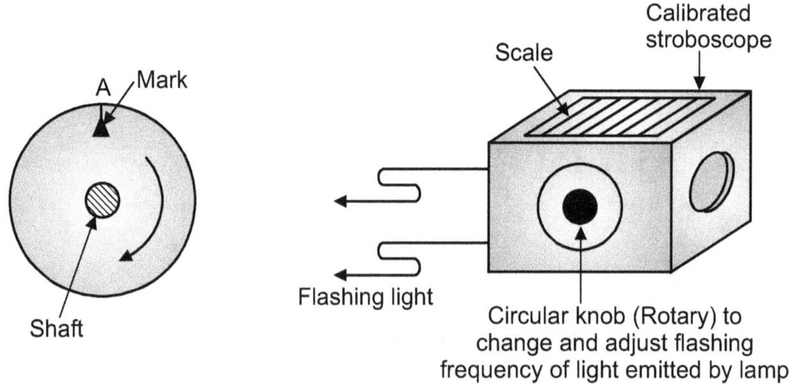

Fig. 4.15: Stroboscope

4.11.4.1 Advantages of Stroboscope

1. It requires no special attachment with the shaft.
2. This method imposes no load on shaft.
3. Contactless operation.

4.11.4.2 Disadvantages of Stroboscope

1. Variable frequency of oscillator cannot be stabilized to give fixed frequency.
2. This method is less accurate. Therefore, it requires the use of digital meters.
3. It cannot be used, where ambient lightening is above a certain value. If it has to be used, then surrounding light must be dim to get effectiveness.

4.11.4.3 Applications of Stroboscope

1. It is a portable and manually operated device used to measure periodic or rotary motions.
2. It is used to measure speed.
3. It is very convenient to use a stroboscope for spot check on machinery speeds and for laboratory work.

4.12 DIFFERENCE BETWEEN CONTACT METHOD OF SPEED MEASUREMENT AND CONTACT-LESS METHOD OF SPEED MEASUREMENT

Sr. No.	Comparative Point	Contact Method of Speed Measurement	Contact-less Method of Speed Measurement
1.	**Physical contact**	Physical contact exists between tachometer and shaft, whose speed is to be measured.	No physical contact exists between tachometer and shaft, whose speed is to be measured.
2.	**Output**	Output is in the form of electrical signal or r.p.m.	Output is in the form of pulses.
3.	**Maintenance**	High maintenance is required due to rotating parts.	Since methods of measurement are contactless, less maintenance is required.
4.	**Applications**	Eddy current or Drag Cup Tachometer, Revolution counter and Timer, Hand speed indicator etc.	Magnetic pick-up type tachometer, Photoelectric tachometer, Stroboscope etc.

4.13 STRAIN MEASUREMENT

- Load carrying capacity of material without excessive deformation or failure is characterized by a term, **"stress"**.
- **Stress** is defined as, **"force experienced per unit area, expressed in pressure units. "**

- Stress cannot be measured directly. Therefore, it is normally calculated from the changes in mechanical dimensions and the applied load.
- **Strain** is measured as **mechanical deformation formed** due to stress with the help of strain gauge measurement devices.
- Linear strain is the ratio of change in length to the original length.
- The **strain gauge element** is defined as, *"a transducer or device used to measure dimensional change on the surface of a structure under test."*
- Strain gauge gives indication of strain at only one point.

4.14 PURPOSE OF STRAIN GAUGES

- Strain gauges are used to measure strains and associated stress.
- Strain gauge serves **two main purposes**.
   1. Strain gauge determines the state of strain existing at a point on a loaded member. This determination is necessary for strain analysis.
   2. Strain gauge acts as a strain sensitive transducer for measuring force, temperature, displacement and acceleration.

4.15 APPLICATIONS OF STRAIN GAUGES

1. Measurement of tension, torque and stress in structures.
2. Measurement of force in load cell.
3. To analyze dynamic strains in complex structures, like bridges, automobiles, roads etc.

4.16 THEORY OF STRAIN GAUGES

- Strain gauges are the metal conductors. If a strip or rod or bar of elastic metal is subjected to tension, its longitudinal (linear) dimension will increase corresponding to reduction in lateral dimensions. This is called as positive strain, where length increases, while its cross-section area decreases.
- We have, $R = \dfrac{\rho \cdot L}{A}$; Therefore, $R \propto L$.
- As the **resistance of a conductor** is directly **proportional to length**, the resistance of gauge wire increases with increase in length, i.e. **positive strain**.
- Increase in resistance of gauge wire is directly proportional to mechanical stress applied on it, therefore strain gauges are called as **positive passive transducers**.

4.17 STRESS-STRAIN RELATIONSHIP

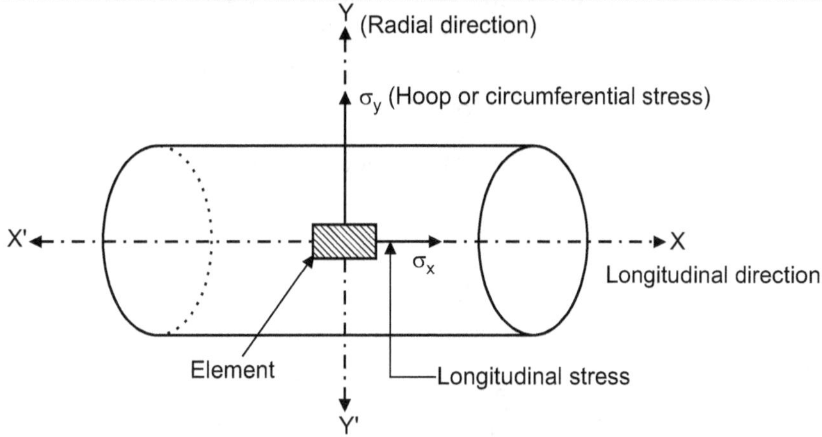

Fig. 4.16: A Component Subjected to Bi-axial Stress System

- By Hooke's law,

$$\text{Stress} = \text{Modulus of elasticity} \times \text{Strain}$$

$$\therefore \quad \sigma = E \times e$$

Strain may be given as,

1. Linear strain = e = $\dfrac{\text{Change in length}}{\text{Total length}} = \dfrac{\delta L}{L}$ (i.e. Longitudinal or Axial strain)

2. Lateral strain = $e_L = \dfrac{\text{Change in diameter}}{\text{Original diameter}} = -\dfrac{\delta d}{d}$

Poisson's ratio gives relation between two strains.

$$\mu = \dfrac{\text{Lateral strain}}{\text{Linear strain}} = \dfrac{e_L}{e}$$

*   Single strain gauges are used, when loading is uni-axial and its direction is known, i.e. when the test member is loaded either in tension or compression.
*   For bi-axial stress system, two or more gauges are used in different directions.

For example:

*   Consider two stresses σ_x and σ_y acting at right angle to each other.
*   With effect of σ_x alone, it will introduce a strain (σ_x/E) in the X-direction and $(-\mu \cdot \sigma_x/E)$ in Y-direction. Similarly, with effect of σ_y alone, it will introduce a strain (σ_y/E) in the Y-direction and $(-\mu \cdot \sigma_y/E)$ in X-direction. Therefore, the resultant strains in X and Y-directions are obtained as,

$$e_x = \frac{1}{E}\,(\sigma_x - \mu\sigma_y)$$

and $$e_y = \frac{1}{E}\,(\sigma_y - \mu\sigma_x)$$

*   Thus, the values of σ_x and σ_y can be computed, if strain values e_x and e_y are determined with the help of strain gauges.

4.18 GAUGE FACTOR

*   Gauge factor is defined as, ***"the ratio of per unit change in resistance to per unit change in length".***

*   Mathematically,   Gauge factor = $\dfrac{\left(\dfrac{\delta R}{R}\right)}{\left(\dfrac{\delta L}{L}\right)}$

*   When axial tensile load is applied on a circular metallic rod, it leads to increase in length of rod with subsequent decrease in cross-sectional aera.

*   If δL = Increase in length of rod   in 'mm'

     L = Original length of rod   in 'mm'

     δd = Decrease in diameter of rod   in 'mm'

     d = Original diameter of rod   in 'mm'

     A = Cross-sectional area of rod = $\frac{\pi}{4}d^2$ in 'mm^2'

then,   Linear strain, e = $\dfrac{\delta L}{L}$

and   Lateral strain, $e_L = -\dfrac{\delta d}{d}$

*   We know that,

   Poisson's ratio, $\mu = \dfrac{e_L}{e}$

$\therefore$ $\mu = \dfrac{\left(-\dfrac{\delta d}{d}\right)}{\left(\dfrac{\delta L}{L}\right)} = -\dfrac{\left(\dfrac{\delta d}{d}\right)}{\left(\dfrac{\delta L}{L}\right)}$

- Resistance is given by,

$$R = \frac{\rho \cdot L}{A}$$

Taking log on both sides,

$$\log_e (R) = \log_e (\rho) + \log_e (L) - \log_e (A)$$

On differentiating, we get,

$$\frac{\delta R}{R} = \frac{\delta \rho}{\rho} + \frac{\delta L}{L} - \frac{\delta A}{A}$$

$$\therefore \quad \frac{\delta R}{R} = \frac{\delta \rho}{\rho} + \frac{\delta L}{L} - \frac{2\delta d}{d} \qquad \left[\because A = \frac{\pi}{4} d^2; \; \therefore \frac{\delta A}{A} = \frac{2\delta d}{d} \right]$$

Dividing the above equation by $\left(\frac{\delta L}{L} \right)$, we get,

$$\therefore \quad \frac{\frac{\delta R}{R}}{\frac{\delta L}{L}} = \frac{\frac{\delta \rho}{\rho}}{\frac{\delta L}{L}} + 1 - 2 \times \frac{\left(\frac{\delta d}{d} \right)}{\left(\frac{\delta L}{L} \right)}$$

$$\therefore \quad \textbf{Gauge factor} = \left(\frac{\frac{\delta \rho}{\rho}}{\frac{\delta L}{L}} \right) + 1 + 2\,\mu \qquad \qquad \ldots (4.1)$$

where, $\quad \dfrac{\delta \rho}{\rho}$ = Change in resistivity due to piezoelectric effect

$\quad\quad 1$ = Change in resistance due to change in length

$\quad\quad 2\mu$ = Change in resistance due to change in cross-sectional area

- A higher gauge factor makes the strain gauge more sensitive. Therefore, electrical output obtained for indication and recording purpose will be greater. Since change in resistivity ($\delta \rho$) is very small, the value of the term $\left(\dfrac{\frac{\delta \rho}{\rho}}{\frac{\delta L}{L}} \right)$ will also be very very small. Therefore, if we neglect it, equation (4.1) becomes,

Gauge factor (G.F.) = 1 + 2μ

4.19 RESISTANCE OR ELECTRICAL STRAIN GAUGES (ELECTRO MECHANICAL TRANSDUCER)

- The principle of electrical strain gauge is based on measurement of change in some electrical characteristics such as, resistance, capacitance or inductance.
- In general, resistance type strain gauges are used.
- It is proportional to strain applied on test specimen.
- It works on principle of change in resistance of wire with change in strain; the change in resistance is measured by using Wheatstone bridge.
- The output of Wheatstone bridge can be calibrated to give strain.

4.19.1 Advantages of Electrical Strain Gauges

1. Simple in construction.
2. Strain gauges can be calibrated in terms of quantities, such as, force, displacement, pressure and acceleration.
3. Very sensitive.
4. Linearity.
5. They provide good output for indicating and recording purposes.
6. Inexpensive and reliable.

4.20 CLASSIFICATION OF RESISTANCE OR ELECTRICAL STRAIN GAUGES

- Resistance or electrical strain gauges are classified according to,
  1. **Bonding:**
     (i) **Bonded strain gauge:**
        - The name "bonded gauges" is given to strain gauges, which are glued to a large structure under stress, i.e. specimen.
     (ii) **Unbonded strain gauge:**
        - Unbonded gauges are cemented.
        - They are simply screwed at desired location, so that, they can be detached or dismantled and can be again attached or assembled for repetitive times, as and when required.
  2. **Construction:**
     (i) Wire type bonded strain gauge
     (ii) Foil type bonded strain gauge
     (iii) Semiconductor type bonded strain gauge

4.21 SELECTION CRITERIA FOR STRAIN GAUGES

- Selection of suitable strain gauge is done by considering,
  1. Size of strain gauge, i.e. availability of space.
  2. Location, where the strain gauge is to be mounted.
  3. High gauge resistance.
  4. Sensitivity.
  5. Environmental conditions, at which, strain measurement is to be done.
  6. Cost
  7. Requirement of protection – Needed or not.
  8. Strain limits of measurements.

4.22 WIRE TYPE BONDED STRAIN GAUGE

Construction:

- Wire type bonded strain gauge consists of a **thin sheet of insulating material** such as, **Paper, Bakelite or Teflon**.
- Generally, strain gauge wire has a diameter 0.025 m. It has uniform cross-sectional area throughout its length.
- **Materials used for metallic wire are,**
  (i) Constantan (Copper – Nickel alloy),
  (ii) Nichrome V (Nickel – Chrome alloy) etc.
- This strain gauge wire is cemented between two thin sheets made-up of insulating material (one at bottom and other at top), in such a way that, spreading of wire permits a uniform distribution of stress. At the time, precaution is taken to protect the strain gauge wire from any mechanical damage. Refer Fig. 4.17.
- Two terminals of wire are taken out as connecting leads or terminal leads, which may be extended using copper wires.
- Such assembled strain gauge is bonded with an adhesive material to the structure under test. This permits a good transfer of strain from the structure under test to the strain gauge wire.

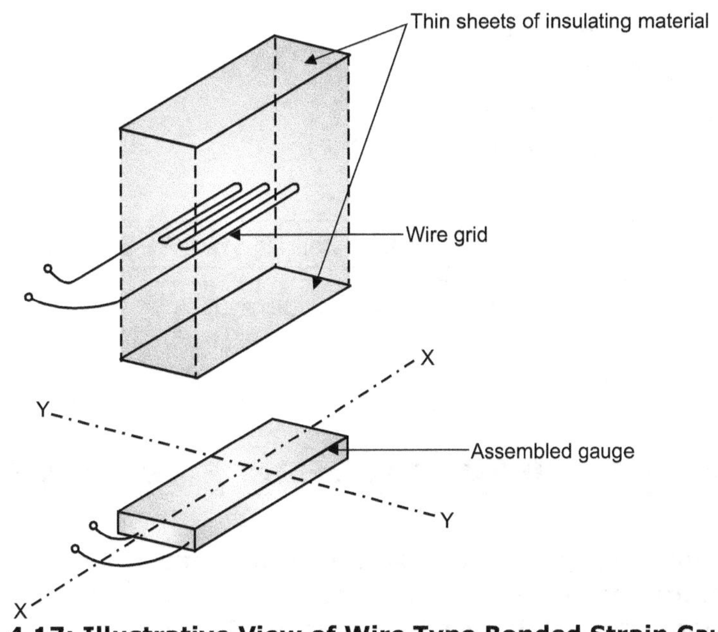

Fig. 4.17: Illustrative View of Wire Type Bonded Strain Gauges

Working:

- Two terminal leads of strain gauge form one arm of Wheatstone's bridge circuit.
- When quantity being measured is zero, the resistances of all arms are so adjusted that the galvanometer shows zero deflection.
- When the strain is applied, resistance of strain gauge wire changes causing unbalance in the bridge. The deflection shown by galvanometer is calibrated in terms of quantity being measured, which may be strain, stress, force, pressure, displacement, acceleration etc. This type of strain gauge is used as load cell.

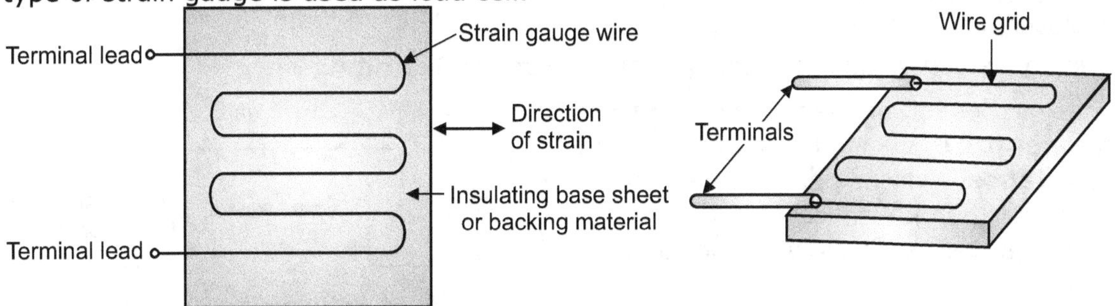

Fig. 4.18: Wire Type Bonded Strain Gauges

4.23 FOIL TYPE BONDED STRAIN GAUGE (METALLIC GAUGES)

- They are similar to wire type strain gauges.
- Only difference is that, metal foils are used in place of strain gauge wires to form grid pattern.
- Foil type resistance gauge has a thickness of 0.005 mm. These gauges are manufactured by printing them on a thin sheet of metal alloy with an acid resistant ink and unprinted portion is etched away.
- Foils provide more surface area, which increases the heat dissipation capacity, so that, they can be used in high temperature area.

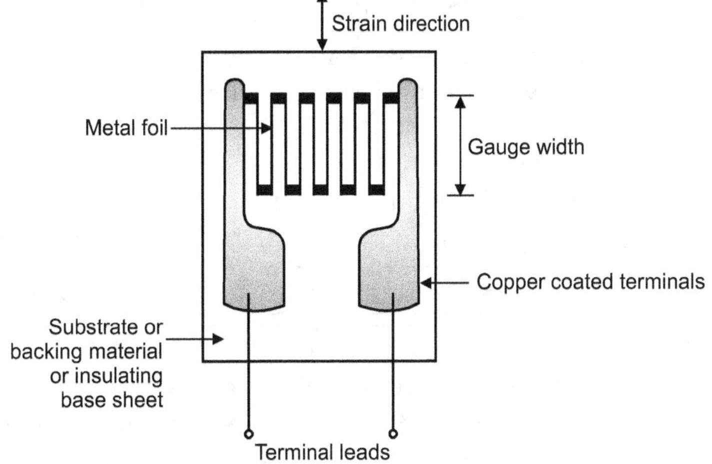

Fig. 4.19: Foil Type Bonded Strain Gauge

- The metals and alloys used for the foil are,
  (a) Constantan (Copper-Nickel alloy),
  (b) Nichrome V (Nickel-Chrome alloy),
  (c) Isoelastic (Ni + Cr + Ma) and
  (d) Platinum.

4.23.1 Advantages of Foil Type Bonded Strain Gauge

1. It has better thermal stability, even at high temperatures.
2. No stress concentration at terminals.
3. Can be fabricated in any size.
4. Low hysteresis.

4.24 SEMICONDUCTOR TYPE BONDED STRAIN GAUGE

Working Principle:

- Semiconductor strain gauge uses the **property of piezo resistivity** possessed by **doped silicon or germaniums.** When such materials are stressed due to load imposed on them causing their deformation, their resistivity changes. This **change in resistivity** (resistance) is used as **a measure of strain**.
- These gauges are produced in the form of wafers of thickness 150 μm.
- The material used may be Silicon or Germanium, in which, exact amount of impurities, such as **Boron** or **Phosphorous** have been **added to import certain desirable charac-teristics**. This process is called as **doping** and the crystals are known as **doped crystals**.

Construction and Working:

- A typical semiconductor strain gauge consists of semiconductor materials and leads enclosed in a protective casing.
- It is bonded on suitable insulating material such as **Teflon.**

- These types of gauges have **negative temperature coefficient of resistance**, which permits very large change in length as compared to other strain gauges, such as wire and foil type gauges.
- **Electrodes** are made up of **gold wires**. They are generally used as one arm of **Wheat-stone's** bridge.

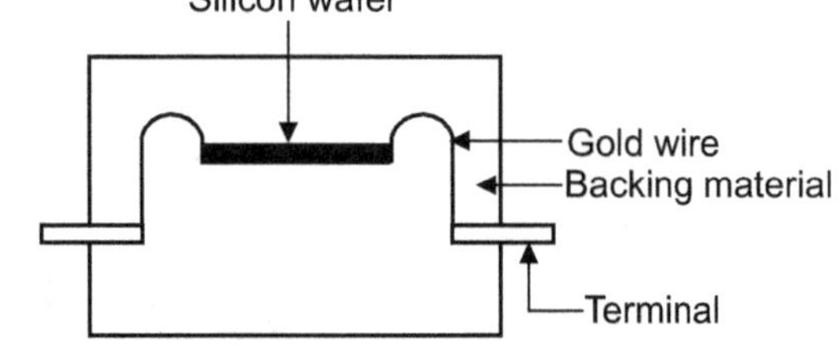

Fig. 4.20: Semiconductor Type Bonded Strain Gauges

4.24.1 Advantages of Semiconductor Type Bonded Strain Gauge

1. Semiconductor strain gauge has high gauge factor (100 to 200) as compared to metallic wires having gauge factor 2. Due to high gauge factor, very small strains or deformations can be measured accurately.
2. Low hysteresis.
3. More life under fatigue loading.
4. Very compact in length from 0.7 to 7 mm, so occupies less space.

4.24.2 Disadvantages of Semiconductor Type Bonded Strain Gauge

1. It is very sensitive to change in temperature, so it is difficult to obtain reliable reading.
2. It is more expensive and difficult to attach to the object under test.
3. It has non-linear characteristic relationship of output voltage with strain.
4. Not suitable for high strain measurements.
5. It can be easily broken as compared to its equivalent wire or foil type strain gauge.

4.24.3 Application of Semiconductor Type Bonded Strain Gauge

1. Use of semiconductor strain gauges is preferred in applications, where high sensitivity is required.

4.25 DIFFERENCE BETWEEN SEMICONDUCTOR GAUGES AND METAL GAUGES

Sr. No.	Comparative Point	Semiconductor Gauges	Metal Gauges
1.	Gauge factor	Very high.	Comparatively small.
2.	Size	Very small.	Large except foil gauges.
3.	Linearity	Poor.	Very good.
4.	Temperature stability	Poor.	Excellent.
5.	Elastic strain range	High.	Low.
6.	Fatigue life	High.	Low.
7.	Protection from strong fluctuating light	Essential due to photo-sensitivity of semiconductor.	Not required.

4.26 APPLICATIONS OF BONDED STRAIN GAUGES

1. For force measurement (For example: Using load cells).
2. For torque measurement (For example: Using strain gauge torsion meter).
3. For finding out maximum stress values.
4. For pressure measurement.

4.27 BACKING MATERIAL

- Backing material is also referred as substrate or base sheet made-up of insulating material.
- Backing material is the portion of the strain gauge, to which, wire or foil grid structure is attached.
- It provides protection to the gauge and helps the grid to retain its original geometric shape after removal of external forces.
- It also acts as electrical insulation for the grid.
- Commonly used strain gauge backing materials are paper, epoxy paper, fibre glass, phenolic (Bakelite) etc.

4.28 BONDING CEMENT

- Bonding cement is the adhesive used to fix the strain gauge on to the test specimen.
- It performs vital function of transmitting the strain from the specimen to the gauge-sensing element.
- Bonding material should be chosen carefully for a particular application.
- Improper bonding of strain gauge to the test specimen can lead to hysteresis, creep, temperature induced zero drift etc.
- Bonding cements are of two types:
  1. Solvent-setting cement is cured by solvent evaporation. For example: Duco cement.
  2. Chemically-reacting cement is cured by polymerization. For example: Phenolic bakelite.

4.29 UNBONDED STRAIN GAUGES

Construction:

- Unbonded strain gauges do not have backing material, so strain is transferred to the resistance wire directly.
- Unbonded strain gauge consists of a fixed frame and a movable frame (armature).
- Movable frame is connected to a rod or bar, which is able to move in either direction under the application of force and strain to be measured.

- Four strain gauge wires are made up of **platinum tungsten alloy** having high tensile strength.
- They have uniform cross-sectional area and equal length and they are connected in such a manner that; one end of each wire is connected to fixed frame and another end is connected to armature or movable frame. Refer wires aa', bb', cc' and dd' in the Fig. 4.21 (a).
- These wires form a Wheatstone's bridge circuit and the galvanometer is calibrated in terms of quantity being measured.

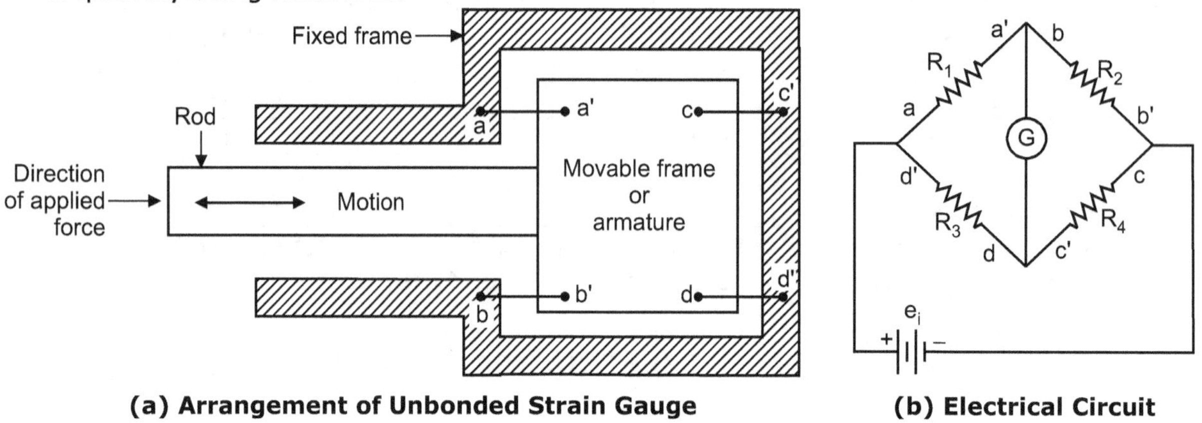

(a) Arrangement of Unbonded Strain Gauge   **(b) Electrical Circuit**

Fig. 4.21: Unbonded Strain Gauge

Working:

- Initially, no force is applied. At this condition of preloading, it is assumed that, all the wires are equally stretched. Since there is no force, there will be no stress and strain. The resistance offered by each strain gauge wire is equal in magnitude. Therefore, the galvanometer shows null position (i.e. zero reading).
- When force to be measured is applied on the rod, then due to applied force, the rod and hence, armature gets moved (say rightwards).
- Due to this movement of armature, the length of aa' and bb' increases (tension), whereas, length of cc' and dd' decreases (compression).
- Due to change in lengths, the resistances of strain gauge wires are changed, which causes unbalancing of Wheatstone's bridge. The galvanometer reading is directly calibrated to read the magnitude of displacement of rod or strain and hence, force applied.

4.29.1 Advantages of Unbonded Strain Gauges

1. Simple in construction.
2. Freedom from faulty insulation.
3. Small creep.
4. Low hysteresis.
5. Due to absence of organic substances, they can be used in high temperature installations.

4.29.2 Disadvantages of Unbonded Strain Gauges

1. High cost.
2. Poor heat dissipation.
3. Very low sensitivity.
4. Resistance of contact lead decreases accuracy.

4.29.3 Applications of Unbonded Strain Gauges

1. It is used for measurement of force, strain etc.
2. It is used for measurement of pressure, displacement etc.
3. In accelerometers.

4.30 DIFFERENCE BETWEEN BONDED TYPE AND UNBONDED TYPE STRAIN GAUGES

Sr. No.	Bonded Type Strain Gauges	Unbonded Type Strain Gauges
1.	The four wires are attached between the moving frame and fixed frame. Therefore, the wires are unbonded.	All the grid-material of wire mesh is embedded in the epoxy thin papers. Therefore, the wires are bonded.
2.	Damage to the wires may take place due to abrupt use, while working.	No any damage to the "Wire Grid Material" during working.
3.	Less life.	More life.
4.	Cannot be easily applied onto the member under test.	Can be easily applied onto the member under test.
5.	Bulky and heavy weight.	Very light weight.

4.31 SELECTION AND INSTALLATION OF STRAIN GAUGE

Selection of Suitable Strain Gauge:
- It is done by considering,
    1. Its size and location, where the strain gauge is to be mounted.
    2. High gauge resistance.
    3. Sensitivity.
    4. Environmental conditions, at which, strain measurement is to be done.
    5. Cost.
    6. Requirement of Protection – Needed or not.
    7. Strain limits of measurements.

Installation procedure:
1. The surface of test specimen should be smooth enough, but not highly polished. Highly polished surface does not provide good adhesion.
2. Surface must be made clean by removing rust, paint and scales (deposited layers of solids or liquids) and other contaminations.
3. For surface preparation, good quality abrasive paper or emery cloth should be used for cleaning.
4. The surface is further cleaned to remove any traces of grease or dirt using solvents like acetone, trichloroethylene with cotton.
5. The strain gauge backing should be clean using suitable solvent to remove oil and grease contamination. After cleaning, strain gauge backing should be dried under electric bulb to remove moisture.
6. A fairly generous layer of recommended cement is spread on prepared clean surface of specimen as well as on strain gauge backing.
7. The gauge is then laid at desired location very carefully, pressed with thumb to exert pressure. Generally, applied pressure and temperature should be around 35 to 180 kPa and 60°C to 200°C respectively.
8. A care must be taken to avoid air gap between gauge and specimen.

4.32 LOAD CELL

- Load cells are elastic devices, which can be used for measurement of force through indirect methods, where force or load or weight is converted into an electric signal, using secondary transducers.
- Load cells utilize an **elastic member** as the **primary transducer** and **strain gauge** as the **secondary transducer.**
- When the combination of the strain gauge and elastic member is used for weighing, it is called as a **"load cell".**

4.33 TYPES OF LOAD CELL

- Various types of load cells are,
  1. Hydraulic load cell
  2. Pneumatic load cell
  3. Strain gauge load cell

4.33.1 Hydraulic Load Cell

- Hydraulic load cell can be used to measure force of very high magnitude (of the order of millions of Newton).
- Hydraulic load cell consists of a closed container filled with oil and covered with diaphragm as shown in Fig. 4.22.
- A pressure measuring device, such as, Bourdon tube pressure gauge is connected to the container.
- When load (P) acts on the diaphragm, it gets deflected and transmits force to the oil. Due to this, pressure is developed in the oil.
- The magnitude of pressure developed is indicated by the pressure gauge, which can be converted into force.
- The pressure gauge can also be directly calibrated in units of force.

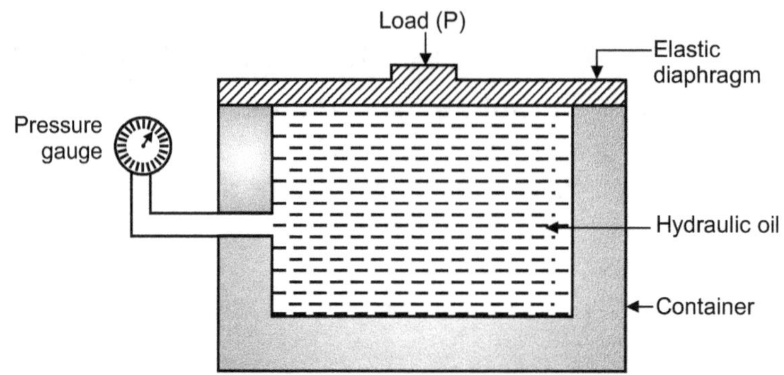

Fig. 4.22: Hydraulic Load Cell

4.33.2 Pneumatic Load Cell

- Pneumatic load cell is used to measure force up to 20,000 N.
- Pneumatic load cell consists of a chamber with a nozzle and covered with elastic diaphragm as shown in Fig. 4.23.
- Initially, pressurised air is supplied through a pipe to the chamber.
- As force or load (P) acts on the diaphragm, it deflects and the gap between diaphragm and nozzle decreases, while pressure P_c in the chamber increases.
- Rise in pressure P_c applies outward force on the diaphragm opposite the force or load P.
- For any force 'P', the diaphragm attains equilibrium under the impact of two equal and opposite forces.
- At this state, chamber pressure is noted using pressure gauge, which can also be directly calibrated in terms of force or load.

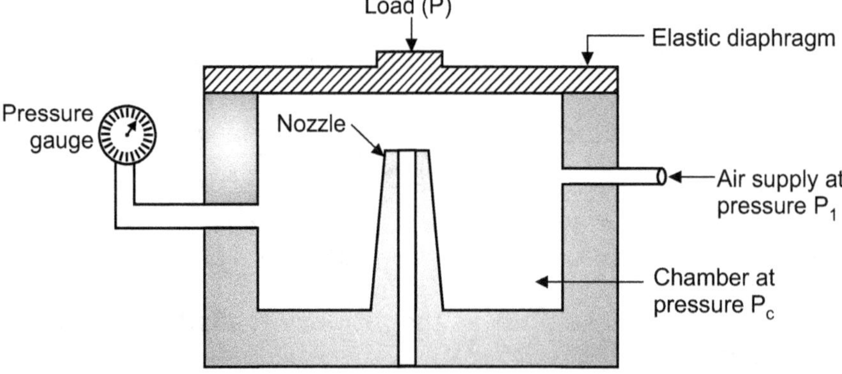

Fig. 4.23: Pneumatic Load Cell

4.33.3 Strain Gauge Load Cell

- Strain gauge Load cell is an application of wire type bonded strain gauge.
- Load cell is an electro-mechanical transducer. These cells convert Load or force (P) into electrical output, which is provided by the strain gauges.
- It translates the change in force or strain into change in voltage.
- This output can be connected to various measuring instruments for indicating, recording and controlling the weight or force.
- Load cells are excellent force-measuring devices, particularly for transient and non-steady forces.

Construction:

- Fig. 4.24 (a) shows a simple strain gauge load cell.
- Load cell consists of a steel cylinder, on which, four identical strain gauges are mounted.
- The gauges R_1 and R_4 are along the direction of applied load (i.e. vertically) and the gauges R_2 and R_3 are attached circumferentially to gauges R_1 and R_4.
- All the four gauges are connected electrically to the four arms of a Wheatstone bridge circuit.

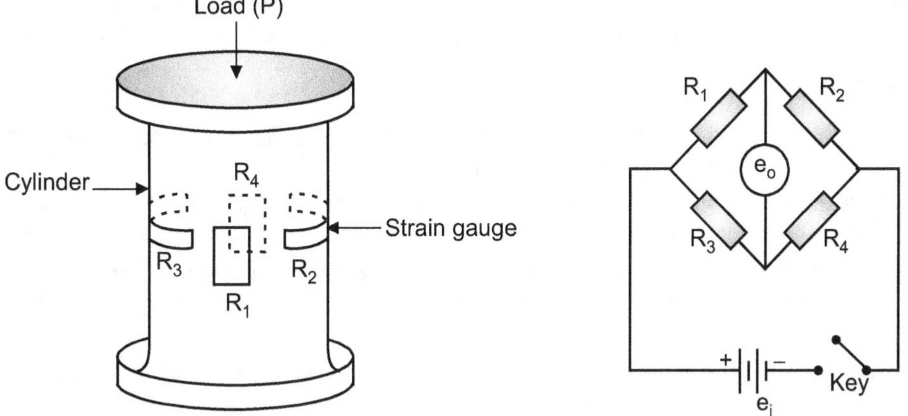

(a) Arrangement of Strain Gauge Load Cell   **(b) Electric Circuit**

Fig. 4.24: Strain Gauge Load Cell

Working:

- When there is no load on the cell, all four strain gauges will have same resistances ($R_1 = R_4 = R_3 = R_2$). It means, Wheatstone's bridge is in balanced condition, and galvanometer shows zero output voltage.
- On the application of a compressive load to the unit, the vertical gauges (R_1 and R_4) undergo compression (i.e. decrease in length causing negative strain). Due to decrease in length, resistance decreases.
- At same time, the circumferential gauges R_2 and R_3 undergo tension (i.e. increase in length causing positive strain). Due to increase in length, resistance increases.
- As a result of change in resistances, Wheatstone's bridge gets unbalanced.
- Thus, an output voltage obtained is a measure of applied load. Refer Fig. 4.24 (b).

4.33.3.1 Advantages of Strain Gauge Load Cell

1. Load cell is small and compact in size.
2. Accuracy is very good.
3. It has good sensitivity and low hysteresis.
4. It is less expensive.

4.33.3.2 Disadvantages of Strain Gauge Load Cell

1. Its performance is affected by non-axial force.
2. It requires temperature compensation network.
3. Excessive stress or force may damage load cell permanently.

4.33.3.3 Applications of Strain Gauge Load Cell

1. Force dynamometer,
2. Crane load monitoring, and
3. Road vehicle weighing devices.

4.34 APPLICATIONS OF LOAD CELLS

1. Scales.
2. Weighbridge or weighing machine for measuring weight of car, truck etc.
3. Force gauges.
4. Torque gauges.
5. Material handling devices such as crane or AGV.
6. Suspension bridges.
7. For measuring press loads.
8. For measuring pressurization force in case of machines, such as, spot welding machines, polishing machines etc.

4.35 ROSETTES

- For uni-axial strain measurement, single strain gauges are used, because its one direction is known.
- **For example:** When the gauge bonded to the specimen is subjected to either tensile or compressive load. Here, we say that member is strained only in one direction, for which,

  linear strain , $e = \dfrac{\sigma}{E}$

- But in actual practice, the member is also subjected to lateral strains (along Y and Z axes). Thus, it becomes a tri-axial strain condition.
- Hence, it is not possible to locate the directions of principal stresses, which makes practically impossible to orient the strain gauge along the directions of principal stresses.
- Therefore, a strain gauge measurement system capable of measuring the values of principal stresses and strains without actually knowing their direction give rise to revolution of rosette.
- For unknown strain directions, a multiple array of gauges i.e. Rosettes are used.
- Gauges are mostly set at 45° to 60° angles between them and with certain calculations; it is possible to know magnitude and direction of principal strains.
- Usually, they have three separate grids with various angular orientation and they can be cemented to the part with no particular attention being paid to the overall gauge orientation.
- The resultant strain on each of the grid is recorded and magnitude and direction of strain is calculated from the collected data.
- Types of rosettes are, (1) Rectangular rosettes, and (2) Delta rosettes.
- The various types of strain rosettes depending upon the arrangement of grids is as shown in Fig. 4.25.

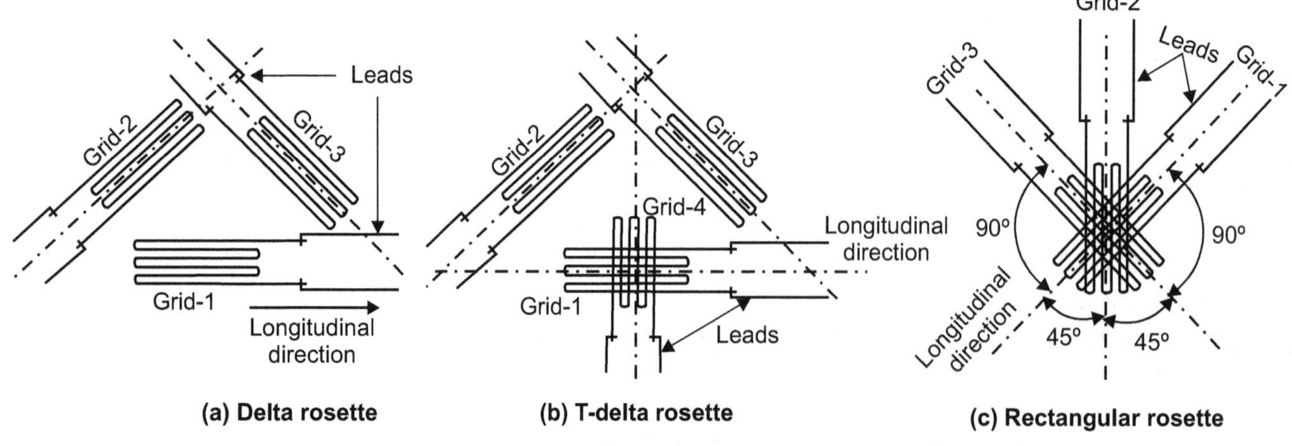

(a) Delta rosette   (b) T-delta rosette   (c) Rectangular rosette

Fig. 4.25

Important:

- While using single element gauge, orientation is a critical factor. So single element gauge is useful only in known direction. But with the use of strain gauge rosettes, there is no restriction on the orientation. They can be used to measure strains, whose magnitude and direction are totally unknown.

- For strain measurement in complex parts, **rosettes** i.e. combination of strain gauges is used instead of single element strain gauges.

- They are available in many combinations, which can be used for strain measurement.

Solved Numerical Type I: "Lathe Tool Dynamometer"

Problem 4.1: *Determine total power consumed by a Lathe for following data.*

(i)   *Spindle speed = 300 r.p.m.*

(ii) *Feed = 0.8 mm/rev.*

(iii) *Mean diameter of job = 100 mm*

(iv) *Tangential force and axis force recorded by Lathe tool dynamometer are 800 N and 1000 N respectively.*

Solution: Given data: N = 300 r.p.m., d = 100 mm = 0.1 m

$$\text{Feed rate, } V = 0.8 \text{ mm/revolution} = 0.8 \times 10^{-3} \text{ m/revolution}$$

$$\text{Tangential force, } F_t = 800 \text{ N}$$

$$\text{Axial force, } F_a = 1000 \text{ N}$$

Procedure:

1. Power consumed in rotating the workpiece:

We have, Torque applied to rotate the workpiece is,

$$T = F_t \times r$$

$$\therefore \quad T = F_t \times \frac{d}{2} = 800 \times \frac{0.1}{2} = \mathbf{40 \ N.m}$$

Therefore, power consumed in rotating the workpiece is,

$$P_R = \frac{2\pi NT}{60} = \frac{2\pi \times 300 \times 40}{60} = \mathbf{1256.63 \ Watts}$$

2. Power consumed in feeding the workpiece:

We have, Power consumed in feeding the workpiece as,

$$P_R = \frac{\text{Axial force} \times \text{Feed rate} \times \text{Number of revolutions per minute}}{60}$$

$$\therefore \quad P_F = \frac{F_a \times V \times N}{60} = \frac{1000 \times 0.8 \times 10^{-3} \times 300}{60} = \mathbf{4 \ Watts}$$

3. Total power consumed by a lathe machine:

Total power consumed by a lathe machine is given by,

$$P = P_R + P_F = 1256.63 + 4 = \mathbf{1260.63 \ N}$$

Solved Numerical Type II: "Inductive Pick-Up Type Contactless Tachometer"

Problem 4.2: *A metallic toothed-rotor having 60 teeth was mounted on a shaft, whose speed was to be determined. The pulses induced in the inductive-type magnetic pick up were registered to be 3600 per second. Estimate shaft speed. If the frequency meter gives count with ± 5% of speed. Calculate the range within which the shaft speed can lie.*

Solution: Given data: Number of teeth = 60, Number of pulses per second = 3600

Procedure:

1. Speed of shaft (N):

We have,   Speed of shaft $= \dfrac{\text{Number of pulses per seconds}}{\text{Number of teeth}}$

$\therefore$ $N = \dfrac{3600}{60}$

$\therefore$ **N = 60 r.p.s.** $= 60 \times 60 =$ **3600 r.p.m.**   ... (1)

2. Range of inductive type magnetic pickup tachometer:

It is given that, frequency range = ± 5% of speed.

Therefore,   ± 5% of speed $= \pm \dfrac{5}{100} \times 3600$

$= \pm$ **180 r.p.m.**   ... (2)

From equations (1) and (2), we get,

Maximum shaft speed = 3600 + 180 = **3780 r.p.m.**

Minimum shaft speed = 3600 − 180 = **3420 r.p.m.**

Thus, range of tachometer is from 3420 r.p.m. to 3780 r.p.m.

Solved Numerical Type III: "Photoelectric Type Contactless Tachometer"

Problem 4.3: *A disc having 60 holes on its periphery is mounted on the shaft. Calculate the speed of the shaft in r.p.m., if an electric (photoelectric) tachometer is used and number of pulses registered are 3000 pulses/sec.*

Solution: Given data: Number of holes, H = 60, Number of pulses per second = 3000

Procedure: Given that, number of pulses per second = 3000

$\therefore$ Speed of shaft $= \dfrac{\text{Number of pulses per second}}{\text{Number of Holes}}$

$\therefore$ $N = \dfrac{3000}{60}$ in r.p.s.

$\therefore$ $N = 50$ in r.p.s.

$\therefore$ $N = 50 \times 60$ in r.p.m.

$\therefore$ **N = 3000 r.p.m.**

Solved Numerical Type IV: "Stroboscope"

Problem 4.4: *A stroboscope projects 6000 flashes per minute on a disk mounted on the shaft of a machine. Find the speed of the machine, if the disk appears stationary and has a single point image of 10 points. What would be the two possible speeds, if the 10 points appear to be revolving once in 12 seconds?*

Solution: Given data: Number of flashes per minute = 6000, Number of marks on disk = 10.

Procedure:

1. Speed of shaft:

We have,   Speed of shaft $= \dfrac{\text{Number of flashes per minute}}{\text{Number of marks on disk}}$

$\therefore$ $N = \dfrac{6000}{10}$ in r.p.m.

$\therefore$ **N = 600 r.p.m.**

2. Two possible shaft speeds:

Since, the marks or points are revolving once in 12 seconds, therefore, marks or points revolving in 60 seconds or a minute will be, $\dfrac{60}{12} = 5$ times in one minute.

$\therefore$ **Slip = 5 r.p.m.**

This slip in speed of shaft may cause the shaft to rotate fastly by 5 r.p.m., or slowly by 5 r.p.m.

$\therefore$ **Slip = ± 5 r.p.m.**

Therefore, **two possible shaft speeds will be,**

$N_{max.}$ = 600 + 5 = **605 r.p.m.**

$N_{min.}$ = 600 − 5 = **595 r.p.m.**

Problem 4.5: *A stroboscope projects 5000 flashes per minute on the disk mounted on the shaft of a machine. Find the speed of the machine, if the disk appears stationary and has a single image of 10 points.*

Solution: Given data: Number of flashes per minute = 5000

Number of marks on disk = 10

Procedure:

We have,   Speed of shaft $= \dfrac{\text{Number of flashes per minute}}{\text{Number of marks on disk}}$

$\therefore$ $\mathbf{N} = \dfrac{5000}{10} = \mathbf{500 \ r.p.m.}$

Problem 4.6: *The speed of turbocharger was measured by stroboscope and for that a radial mark was made on the rotating shaft. The synchronization as attained for the highest rate of flashing and subsequently the flashing rate was reduced and a single image observed at reduced flash rates. Calculate the speed of turbocharger, if synchronization is achieved for stroboscope setting of 3600, 1800, 1200, 900 and 720 r.p.m.*

Solution: Given data: Number of flashing frequency, m = 5

Highest flashing frequency = $f_{max.}$ = 3600 r.p.m.

Lowest flashing frequency = $f_{min.}$ = 720 r.p.m.

Procedure:

We have,   Speed of turbocharger $= \dfrac{f_{max} \cdot f_{min.} \cdot (m-1)}{f_{max.} - f_{min.}}$

$\therefore$ $N = \dfrac{3600 \times 720 \times (5-1)}{3600 - 720}$

$\therefore$ $N = \mathbf{3600 \ r.p.m.}$

Solved Numerical Type V: "Strain Measurement"

Problem 4.7: *Strain gauge is bonded to a beam 0.1 m long and has a cross-sectional area 4 cm². Young's modulus for steel is 207 GN/m². The gauge has an unstrained resistance of 240 ohms and a gauge factor of 2.2. When the load is applied, the resistance of gauge changes by 0.013 ohms. Calculate the change in length of the steel beam and the amount of force applied to the beam.*

Solution: Given data: $L = 0.1 \ m$, $\mathbf{A} = 4 \ cm^2 = \mathbf{4 \times 10^{-4} \ m^2}$

$\mathbf{E} = 207 \ GN/m^2 = \mathbf{207 \times 10^9 \ \dfrac{N}{m^2}}$

$R = 240 \ \Omega$, Gauge factor (G.F.) = 2.2, $\delta R = 0.013 \ \Omega$

Procedure:

1. Change in length:

We have,   Gauge factor $= \dfrac{\left(\dfrac{\delta R}{R}\right)}{\left(\dfrac{\delta L}{L}\right)}$

$\therefore$ $2.2 = \dfrac{\left(\dfrac{0.013}{240}\right)}{\left(\dfrac{\delta L}{0.1}\right)}$

$\therefore$ $2.2 = \dfrac{0.013}{240} \times \dfrac{0.1}{\delta L}$

$\therefore$ $\delta L = \mathbf{2.46 \times 10^{-6} \ m}$

2. Amount of force applied to the beam:

We have, $\text{Strain} = \dfrac{\text{Change in length}}{\text{Original length}}$

$\therefore$ $e = \dfrac{\delta L}{L} = \dfrac{2.46 \times 10^{-6}}{0.1} = \mathbf{2.46 \times 10^{-5}}$

We have that,

$$\text{Young's modulus} = \frac{\text{Stress}}{\text{Strain}}$$

$$\therefore \quad E = \frac{\sigma}{e}$$

$$\therefore \quad \sigma = E \times e = 207 \times 10^3 \times 2.46 \times 10^{-5}$$

$$\therefore \quad \boldsymbol{\sigma = 5092.2 \times 10^3 \ N/m^2}$$

We have,

$$\text{Stress} = \frac{\text{Load}}{\text{Area}}$$

$$\therefore \quad \sigma = \frac{W}{A}$$

$$\therefore \quad \boldsymbol{W = \sigma \times A = 5092.2 \times 10^3 \times 4 \times 10^{-4} = 2036.81 \ N}$$

Problem 4.8: *A resistance strain gauge is used for strain measurement. The gauge factor is 4.2. Calculate Poisson's ratio, neglecting piezo-resistive effects.*

Solution: Given data: Gauge factor = 4.2

Procedure:

We have, $\quad$ Gauge factor $= 1 + 2\mu$

$$\therefore \quad 4.2 = 1 + 2\mu$$

$$\therefore \quad \mu = \boldsymbol{1.6}$$

Thus, Poisson's ratio, $\quad \mu = 1.6$

Problem 4.9: *A 100 ohms strain gauge is bonded to a low carbon steel bar, which has been subjected to a tensile load. The bar has a preload uniform cross-sectional area of $0.5 \times 10^{-4} \ m^2$ and Young's modulus for low carbon steel is 200 GN/m^2. If a load of 50 kN produces a change of 1 ohm in the gauge resistance, determine the gauge factor for the strain gauge.*

Solution: Given data: $\quad R = 100 \ \Omega, \ A = 0.5 \times 10^{-4} \ m^2$

$$\boldsymbol{E = 200 \ GN/m^2 = 200 \times 10^9 \ N/m^2}$$

$$\textbf{Load, W} = 50 \ kN = \boldsymbol{50 \times 10^3 N}$$

$$\delta R = 1\Omega$$

Procedure:

We have, $\quad$ Young's modulus $= \dfrac{\text{Stress}}{\text{Strain}}$

$$\therefore \quad E = \frac{\sigma}{e} = \frac{\left(\dfrac{\text{Load}}{\text{Area}}\right)}{\left(\dfrac{\text{Change in length}}{\text{Original length}}\right)}$$

$$\therefore \quad E = \frac{\left(\dfrac{W}{A}\right)}{\left(\dfrac{\sigma L}{L}\right)} = \frac{W}{A} \times \frac{L}{\delta L}$$

$$\therefore \quad 200 \times 10^9 = \frac{50 \times 10^3}{0.5 \times 10^{-4}} \times \frac{L}{\delta L}$$

$$\therefore \quad \frac{\delta L}{L} = \frac{50 \times 10^3}{200 \times 10^9 \times 0.5 \times 10^{-4}} = \boldsymbol{5 \times 10^{-3}} \qquad \dots (1)$$

We have, Gauge factor (G.F.) $= \dfrac{\left(\dfrac{\delta R}{R}\right)}{\left(\dfrac{\delta L}{L}\right)}$

$$\therefore \quad \text{Gauge factor (G.F.)} = \frac{\dfrac{1}{100}}{5 \times 10^{-3}} \qquad \text{[From equation (1)]}$$

$$\therefore \quad \text{Gauge factor (G.F.)} = \boldsymbol{2}$$

Problem 4.10: *A metallic electric resistance strain gauge of 120 ohm resistance with gauge factor of 2 is cemented onto an aluminium test surface. If the yield stress is 200 $\times$ 10^6 N/m^2 and Young's modulus of elasticity is 700 $\times$ 10^9 N/m^2, determine the change in resistance of the gauge that would be caused by loading the material up to yield point.*

Solution: Given data: R = 120 Ω, Gauge factor = 2

$$\sigma = 200 \times 10^6 \text{ N/m}^2$$

$$E = 700 \times 10^9 \text{ N/m}^2$$

Procedure:

We have,   Young's modulus $= \dfrac{\text{Stress}}{\text{Strain}}$

$\therefore$ $E = \dfrac{\sigma}{e} = \dfrac{\sigma}{\left(\dfrac{\delta L}{L}\right)}$

$\therefore$ $\dfrac{\delta L}{L} = \dfrac{\sigma}{E} = \dfrac{200 \times 10^6}{700 \times 10^9} = \mathbf{2.86 \times 10^{-4}}$ ... (1)

We have,   Gauge factor, G.F. $= \dfrac{\left(\dfrac{\delta R}{R}\right)}{\left(\dfrac{\delta L}{L}\right)}$

$\therefore$ $2 = \dfrac{\left(\dfrac{\delta R}{120}\right)}{2.86 \times 10^{-4}}$ [From equation (1)]

$\therefore$ $\delta R = \mathbf{0.069\ \Omega}$

- **Dynamometer** is a device used to measure the torque being exerted along a rotating shaft and hence, power transmitted by shaft.
- **Dynamometers** are classified as, **Absorption** type dynamometer, **Transmission** dynamometer and **Driving** dynamometer.
- Eddy Current Dynamometer works on the principle that, *"when a conductor moves through a magnetic field, it induces eddy currents (called as local currents). These eddy currents flow in a short circular path around the conductor and are dissipated in the form of heat."*
- **In Strain Gauge Transmission Dynamometer**, four bonded-wire strain gauges are mounted at an angle of 45° with respect to the axis of rotation. They are placed in **pairs, but diametrically opposite.** Further, these strain gauges are connected to Wheatstone bridge circuit. The output of Wheatstone bridge is proportional to torsion and hence, to applied torque on shaft.
- **Tachometer** is defined as, "an instrument used to measure the angular velocity of shaft and express it in number of revolutions made per minute (in r.p.m.)."
- **Slipping clutch tachometer** consists of a driving shaft and an indicator shaft, connected by means of slipping clutch.
- **Hand speed indicator** consists of a spindle having conical tip. Therefore, spindle is also called as conical spindle. Due to conical tip, spindle can be connected or attached to the shaft, whose angular speed is to be measured.
- **Working principle of centrifugal tachometer:** When a body revolves about an axis, it experiences a radially outward force. This force is called as **centrifugal force**. This centrifugal force is a function of rotational speed. Therefore, it is used to measure instantaneous speed by means of centrifugal force.
- **Electrical tachometer** converts the rotational speed into electrical signal, which can be measured and calibrated to give rotational speed directly. In short, it is a transducer, which produces an electrical signal in proportion to the rotational speed.
- In **Eddy current or Drag type of tachometer**, the transducer produces an analog signal in the form of continuous drag due to eddy current, induced in the cup.

- In **contact-less type of tachometer**, there is no direct contact with the shaft, whose speed is to be measured. The tachometer produces pulse from a rotating shaft.
- **Inductive pick-up type tachometer** consists of a small permanent magnet with a coil wound on it. This permanent magnet is also called as **magnetic pick-up**.
- In **Capacitive pick-up tachometer,** change in capacitance can be calibrated in terms of angular speed of shaft on readout device.
- **Photoelectric tachometer** uses an opaque disc having equidistant holes on its periphery. This disc is mounted on a rotating shaft, whose speed is to be measured.
- **Working principle of stroboscope:** "If a strong light is caused to flash on the moving object, the object will appear stationary, if the flashing frequency is equal to speed of moving object." In other words, "To synchronize a flashing light with rotation of shaft, making it appears as standstill (stationary)."
- **Strain** is measured as **mechanical deformation formed** due to stress with the help of strain gauge measurement devices.
- The **strain gauge element** is defined as, **"a transducer or device used to measure dimensional change on the surface of a structure under test."** Strain gauge gives indication of strain at only one point.
- **Purpose of Strain Gauge:**
  (i) Strain gauge determines the state of strain existing at a point on a loaded member. This determination is necessary for strain analysis.
  (ii) Strain gauge acts as a strain sensitive transducer for measuring force, temperature, displacement and acceleration.
- **Wire type bonded strain gauge** consists of a thin sheet of insulating material such as, paper, Bakelite or Teflon.
- In **foil type bonded strain gauge**, metal foils are used in place of strain gauge wires to form grid pattern.
- **Working principle of semiconductor strain gauge**: It uses the **property of piezoresistivity** possessed by **doped silicon or germaniums.** It has **negative temperature coefficient of resistance.**
- **A typical semiconductor strain gauge** consists of semiconductor materials and leads enclosed in a protective casing.
- **Backing material (Substrate)** is the portion of the strain gauge, to which, wire or foil grid structure is attached. It provides protection to the gauge and helps the grid to retain its original geometric shape after removal of external forces. It also acts as electrical insulation for the grid.
- Commonly used strain gauge **backing material** are **paper, epoxy paper, fibre glass, phenolic (Bakelite) etc.**
- **Bonding cement** is the adhesive used to fix the strain gauge on to the test specimen.
- **Un-bonded strain gauges** do not have backing material, so strain is transferred to the resistance wire directly.
- **Load cells** are elastic devices that can be used for measurement of force through indirect methods through use of secondary transducers. Load cells utilize an elastic member as the primary transducer and strain gauges as secondary transducer.
- Load cells are excellent force-measuring devices, particularly for transient and non-steady forces. Load cells are used in **Scales, Weighbridge, Force gauges, Torque gauges etc**.
- Various types of load cells are, (i) Hydraulic load cell, (ii) Pneumatic load cell, (iii) Strain gauge load cell.
- **Hydraulic load cell** is used to measure force of very high magnitude (of the order of millions of Newton). Hydraulic load cell consists of a closed container filled with oil and covered with diaphragm.
- **Pneumatic load cell** is used to measure force up to 20,000 N. Pneumatic load cell consists of a chamber with a nozzle and covered with elastic diaphragm.
- **Strain gauge load cell** is an application of wire type bonded strain gauge. This cell converts load or force (P) into electrical output, which is provided by the strain gauges.

Theory Questions For Practice

1. Define Dynamometer. State its types and application of each.
2. Explain with neat sketch, strain gauge transmission dynamometer.
3. Explain working principle of eddy-current dynamometer.

4. Explain working of eddy current dynamometer with a neat labelled diagram.

5. List speed measurement devices. Which device is used to measure the speed of automobile?

6. Explain the difference between contact type and contactless type speed measurement devices. State two examples of each type.

7. Explain the working of electrical tachometer.

8. What is stroboscope? Explain its working principle.

9. Explain the working of stroboscope with neat sketch.

10. Explain construction and working of inductive pick-up tachometer.

11. Explain working principle of drag cup type tachometer with neat sketch.

12. Explain with neat sketch, the working of slipping clutch tachometer.

13. Draw neat sketch of slipping clutch tachometer and state its working principle.

14. Explain strain gauge.

15. Define strain gauge. List the types of strain gauges.

16. Enlist different types of strain gauges? Explain.

17. Explain the construction and working of bonded strain gauge with the help of diagram.

18. Explain working and application of bonded strain gauge.

19. Draw and explain the semiconductor strain gauge. List its advantages.

20. List the main requirements of material used for a strain gauge.

21. What are materials used for strain gauge?

22. Explain with neat sketch, working of hydraulic load cell.

23. Draw neat sketch of pneumatic load cell and explain its working principle.

24. Explain with neat sketch, how strain gauge load cell is used?

25. Give advantages and disadvantages of strain gauge load cell.

26. State any two applications of strain gauge load cell.

Numerical Problems For Practice

1. Determine total power consumed by a Lathe for following data.

    (i)   Spindle speed = 600 r.p.m.

    (ii) Feed = 0.16 mm/rev.

    (iii) Mean diameter of job = 200 mm

    (iv) Tangential force and axis force recorded by Lathe tool dynamometer are 1600 N and 2000 N respectively.

2. A metallic toothed-rotor having 120 teeth was mounted on a shaft, whose speed was to be determined. The pulses induced in the inductive-type magnetic pick up were registered to be 7200 per second. Estimate shaft speed. If the frequency meter gives count with $\pm$ 5 % of speed. Calculate the range within which the shaft speed can lie.

3. A disc having 90 holes on its periphery is mounted on the shaft. Calculate the speed of the shaft in r.p.m., if an electric (photoelectric) tachometer is used and number of pulses registered are 4500 pulses / sec.

4. A stroboscope projects 7500 flashes per minute on the disk mounted on the shaft of a machine. Find the speed of the machine, if the disk appears stationary and has a single image of 15 points.

5. Strain gauge is bonded to a beam 0.2 m long and has a cross-sectional area 8 cm^2. Young's modulus for steel is 207 GN/m^2. The gauge has an unstrained resistance of 480 Ohms and a gauge factor of 2.2. When the load is applied, the resistance of gauge changes by 0.026 Ohms. Calculate the change in length of the steel beam and the amount of force applied to the beam.

6. A resistance strain gauge is used for strain measurement. The gauge factor is 3.2. Calculate Poisson's ratio, neglecting piezo-resistive effects.

7. A 150 Ohms strain gauge is bonded to a low carbon steel bar, which has been subjected to a tensile load. The bar has a preload uniform cross-sectional area of 0.1×10^{-5} m^2 and Young's modulus for low carbon steel is 200 GN/m^2. If a load of 100 kN produces a change of 2 Ohms in the gauge resistance, determine the gauge factor for the strain gauge.

■■■

Chapter 5

AUTOMATION

5.1 INTRODUCTION TO MACHINE TOOL AUTOMATION

- Every industry has an endeavor to improve the quality of the finished product. This objective has forced the industry to convert the manual operations into mechanical activities. (i.e. to replace the human by machine.)
- Simultaneously it also aims to manufacture the product in the most economical manner.
- This has led to the use of two words: mechanization and automation.

5.1.1 Mechanization

- The term mechanization means that the operations are carried out by machines instead of being performed by human. But the process is monitored and controlled by human only.
- The movement of tool and work piece is automatic and the tool cycle is automatic. But the operations like loading, unloading, clamping, checking the work piece dimensions are done manually by the operator.
- In mechanization there is no provision for feedback, hence it is called as open loop system.
- Mechanization refers to semi-automatic machines.
- It is a primary stage of automation.

  For example,
  (i) In earlier days weaving was done by hand, now it is done on machines.
  (ii) Tractors have mechanized the work of farmer.

5.1.2 Automation

- The term automation means a higher degree of mechanization.
- It represents a process in which all the operations like material handling, material processing and inspection are performed automatically.
- It means that operator is required only for supervision and can operate number of machines at one time.
- Automation can be defined as, "a technology concerned with the application of mechanical, electronic and computer based systems to operate and control the production".
- Automation is the creation and application of technology to monitor and control the production and delivery of products and services. In today's world for any industry to survive in the competitive market, it must go for automation.
- It is implemented by using a program of instruction combined with a control system that executes the program.
- Automation means a closed loop system in which there is a provision of feedback.
- Because of the growing ubiquity of automation, any categorization of automated tasks and processes is incomplete.
- Automation leads to increased productivity, product quality, safety and reduce manual/periodic inspection, production cost and to be operator friendly.
- For example,
  (i) Assembly line of an automobile.
  (ii) An automated sugar factory.
  (iii) Printing of newspaper.

5.2 NEED OF AUTOMATION

- Although the basic reason for automation is quality of product, higher production rates and an economical product but the reasons for automation vary from the manufacturing industry to industry.
- Some of the reasons to justify automation are:
  - Increase labour productivity.
  - Reduce labour cost.
  - Mitigate the effects of labour shortages.
  - Reduce or eliminate routine manual and clerical tasks.
  - Improve worker safety.
  - Improve product quality.
  - Reduce manufacturing lead time.
  - Accomplish processes that cannot be done manually.
  - Reduce unit cost.

5.3 BASIC ELEMENTS OF AUTOMATED SYSTEM

- An automated system is used to operate some process. There may be various elements in any automated system but the basic elements of an automated system are shown in Fig. 5.1.

  (i) Power - to accomplish the process and operate the automated system.

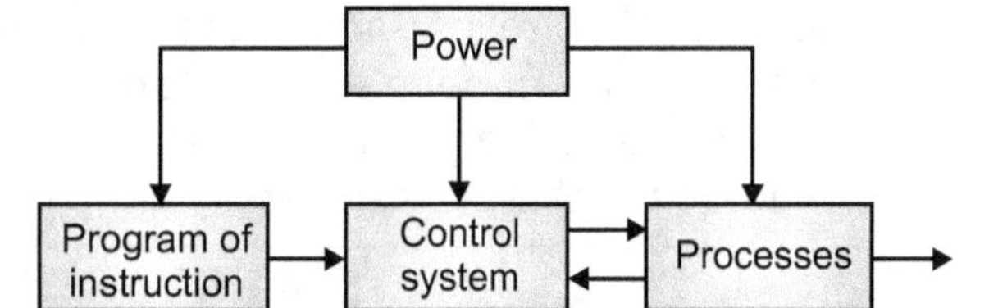

Fig. 5.1: Basic Elements of Automated System

  (ii) Program of instructions – to direct the process.

  (iii) Control system – to actuate the instructions.

(i) Power:

- Power is required to drive the elements as well as the controls. The principal source of power in automated systems is electricity. Electrical power is widely available and is an important part of any industrial infrastructure. It can be readily converted into alternative energy forms like mechanical, thermal, light, acoustic, hydraulic, and pneumatic. Renewable source of power include fossil fuels, solar energy, water and wind.

- However, their exclusive use is uncommon in automated systems. In many cases when alternative power sources are used to drive the process itself, electrical power is used for the controls that automate the operation.

- For example, in casting or heat treatment, the furnace may be heated by unconventional source of energy, but control system to regulate temperature and time cycle is electrical.

- Power is required for the process to drive the process itself, to load and unload the work unit, to facilitate material movement between operations, for control units to automate, Power to actuate the control signals, data acquisition and information processing.

(ii) Program of Instructions:

- Set of commands that specify the sequence of steps in the work cycle and the details of each step are called as program of instruction.

- For example, in CNC part program, in order to set temperature setting of a furnace, to switch motor ON or OFF, there are one or more activities involving changes in one or more process parameters.

- Modern controllers used in automated systems are based on digital computers. Instead of cams, timers, relays, and other hardware devices, the programs for computer controlled equipment are contained in magnetic tape, diskettes, compact disks (CDROMs), computer memory and other modern storage technologies. Virtually all new equipment that perform the above mass production operations are designed with some type of computer controller to execute their respective processing cycles. The use of digital computers in the process controller allows improvements and upgrades to be made.

(ii) Control System:

- o The control element of the automated system executes the program of instructions. The control system causes the process to accomplish its defined function. It may be to carry out some manufacturing operation, give out money from ATM, vary the temperature of an AC, spot welding of car body by robot arm.

- o The controls in an automated system can be either closed loop or open loop. A closed loop control system, also known as a feedback control system is one in which the output variable is compared with an input parameter and any difference between the two is used to drive the output into agreement with the input.

- o In contrast to the closed loop control system, an open loop control system operates without the feedback loop. In this case, the controls operate without measuring the output variable, so no comparison is made between the actual value of the output and the desired input parameter.

- o The controller relies on an accurate model of the effect of its actuator on the process variable. With an open loop system, there is always the risk that the actuator will not have the intended effect on the process, and that is the disadvantage of an open loop system. Its advantage is that, it is generally simpler and less expensive than a closed loop system.

5.4 LOW COST AUTOMATION

- • The lay man meaning of Low Cost Automation is to implement an automation solution at the lowest cost. LCA is defined as 'a system where man and machine collaborate and work together, to achieve a goal with less investment'.

- • An automation, which is easy to implement with high flexibility and reliability, which occupies less space (lean concept), that needs zero or minimum maintenance (high efficiency), with minimum investment and running cost, is termed as LCA.

- • Low cost automation (LCA) is one solution especially for medium and small scale industries who find it difficult to implement automation.

- • LCA involves the introduction of standard equipment, mechanisms and devices to convert manual operations to automatic ones by making use of parts or sub-assemblies of old unutilized machines, mechanisms, systems which are available free or at very lower cost.

- • The steps to achieve LCA are described below:
  - o Design the machine, system, mechanism as per the requirement.
  - o Decide upon the parts, sub-assemblies, sub-systems required.
  - o Check for the above requirements from old unutilized machines, mechanisms, systems, if they are available.
  - o Inspect the parts, sub-assemblies, subsystems available and see that they meet the design requirements.
  - o Use them for developing new machine, system, mechanism.
  - o Due to use of parts of old unutilized machines, mechanisms, systems the overall cost of development of new systems will be very low. Thus, automation is possible at lower cost.

- • Common areas in the manufacturing process for LCA application are machining, cold extrusion, grinding, material handling, quality inspection, dimensional accuracy, surface finishing and assembly and packaging. Besides, areas like storage systems, Handling Systems, Assembly Lines, Production Lines, Production Cells, Machines, Computers, Controllers, Software etc. can also be considered for LCA.

5.5 EXAMPLES OF AUTOMATION

- The examples of automation are numerous as almost all services are using automation whether it is manufacturing industry, service industry, business orgamization or support system.
- Few examples of automation are described below:
  - (a) Self driven trucks that makes long distance deliveries without a driver.
  - (b) Self-driving vehicles where the vehicle itself is a driver which can take you safe at the desired location.
  - (c) Machine automation in packaging of products, pharmaceutical industry, milk bottling plant, assembly of various components etc.
  - (d) Use of robots in transfer of materials in hospitals, cleaning of passages.
  - (e) Robotic arms for assembly of parts, pick and drop, cleaning of manholes etc.
  - (f) Automatic storage and retrieval system to deliver exact material in right quantity from the warehouse.
  - (g) Home appliances like fully automatic washing machine, toaster, coffee makers, automatic furniture etc.
  - (h) Office automation which uses information technology to automate business processes.
  - (i) Online payment systems through use of applications like Phone Pe, BHIM, Ulip etc.
  - (j) Alexa system used for voice interaction, music playback, making to-do lists, setting alarms, playing audiobooks, and providing weather, traffic, sports, and other real-time information,
  - (k) Full automation in sugar factory, newspaper printing, pencil manufacturing etc.

5.6 TYPES/LEVELS OF AUTOMATION

- The automation system replaces the conventional hard-wired relay system by automating the process functions with the use of various hardware and software tools with minimal human intervention.
- There are different types of automation systems used in different applications, but all automated systems are not similar.
- Broad classification of the types of automation is discussed below in brief.
- Automated manufacturing system can be classified into the following types:
  - (i) Fixed automation
  - (ii) Programmable automation
  - (iii) Flexible automation
  - (iv) Integrated automation

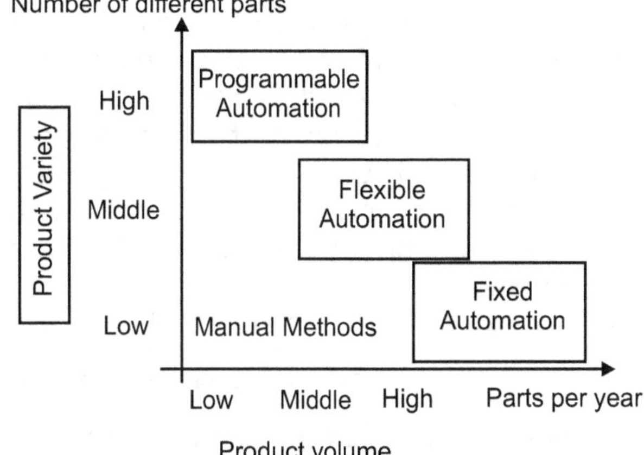

Fig. 5.2: Types of Automation

- The relative positions of the three types of automation for different production volumes and product varieties are represented in the Fig. 5.2.

5.7 FIXED AUTOMATION

- Fixed automation also referred to as hard automation, is the use of special purpose equipment to automate an assembly line.
- Each of the operation in the sequence is usually simple, involving perhaps a plain linear or rotational motion or an uncomplicated combination of two.
- The machines are arranged as per the sequence of operations to be performed.
- It is inflexible and cannot accommodate the variety of product.
- The product design is constant for a long period of time.
- The parts are produced on a large scale.

- The production rates are very high.
- The initial set-up cost of unit is high but the production of large units makes the unit cost attractive.
- The economic justification for fixed automation is found in products with very high demand rates, high volumes and low variability work.
- For example, Transfer lines, automated assembly lines etc.

5.7.1 Features of Fixed Automation

- High initial investment for custom–Engineered equipment.
- High production rates.
- Relatively inflexible in accommodating product changes.

5.7.2 Advantages of Fixed Automation

- Low unit cost.
- Automated material handling.
- High production rate.

5.7.3 Disadvantages of Fixed Automation

- High initial investment.
- Relatively inflexible in accommodating product changes.
- Not designed to accommodate a variety of tasks or setups.
- Modifying hard automation can be costly and time-consuming.

5.8 PROGRAMMABLE AUTOMATION

- Programmable automation also referred to as soft automation, includes equipment that has been designed to accommodate a variety of product configurations.
- In programmable automation, the production equipment is designed with the capability to change the sequence of operations to accommodate different product configurations.
- The sequence of operation depends on the program written.
- A new program has to be written for a new product.
- Flexible to deal with variations and changes in product configuration.
- The physical set-up of the machine needs to be changed for a new product.
- It is used for low volume production where parts are produced in batches.
- High investment in general purpose equipment.
- The economic justification for fixed automation is found in products with low and medium production.
- For example, NC machine tool, PLC, industrial robots etc.

5.8.1 Features of Programmable Automation

- Some of the features that characterize programmable automation are:
  - High investment in general-purpose equipment.
  - Low production rates relative to fixed automation.
  - Flexibility to deal with changes in product configuration.
  - Most suitable for batch production.

5.8.2 Advantages of Programmable Automation

- Flexible to deal with design variations.
- Suitable for batch production.

5.8.3 Disadvantages of Programmable Automation

- High investment in general purpose equipment.
- Lower production rate than fixed automation.
- New program is to be written for every new batch.

5.9 FLEXIBLE AUTOMATION

- A flexible automation system is capable of producing a variety of parts with virtually no time lost for changeovers from one part style to the next. **Flexible automation is an extension of programmable automation:**
  - o Capable of producing variety of parts.
  - o No lost of time for changeover when the product's design changes.
  - o High initial investment.
  - o Continuous type of production.
  - o Production rate is medium.
  - o Flexible to deal with product design variations.
  - o The economic justification for fixed automation is found in production of high mix of products that are continually changing or developing. It can handle both high- and low-volume production with ease.
  - o For example, Flexible manufacturing system etc.

5.9.1 Features of Flexible Automation

- The features of flexible automation can be summarized as follows:
  1. High investment for a custom-engineered system.
  2. Continuous production of variable mixtures of products.
  3. Medium production rates.
- Flexibility to deal with product design variations.

5.9.2 Advantages of Flexible Automation

1. Continuous production of variable mixtures of product.
2. Flexible to deal with product design variation.

5.9.3 Disadvantages of Flexible Automation

1. Medium production rate
2. High investment.
3. High unit cost relative to fixed automation.

5.10 INTEGRATED AUTOMATION

- Manufacturing functions are linked through an integrated computer network.
- These manufacturing functions include production planning and control, shop floor control, quality control, purchasing, marketing etc.
- Allow changes in product design to reduce costs and to optimize production requirements.
- For example, Computer integrated manufacturing (CIM).

5.11 BENEFITS OF AUTOMATION

(i) Increase in productivity.
(ii) Human fatigue is greatly minimised.
(iii) Good quality of product is obtained.
(iv) One operator can operate more than one machine simultaneously.
(v) The components produced are uniform.
(vi) Fewer inventories are required due to use of standard parts.
(vii) Less floor space area is required.

5.12 ADVANCED AUTOMATION FUNCTION

- Apart from providing automated instructions and controlling the system, an automated system should be capable of executing advanced functions that are not specific to a particular work unit. These advanced functions are concerned with enhancing the performance and safety of the equipment.
- Advanced automation functions include the following:
  - o Safety monitoring.
  - o Maintenance and repair diagnostics.
  - o Error detection and recovery.

5.12.1 Safety Monitoring

- One of the reasons of automation in a manufacturing operation is to remove workers from a hazardous working environment. Another reason of safety in automation is to protect the equipment associated with the system.

- An automated system is often installed to perform a potentially dangerous operation that would otherwise be accomplished manually by human workers. However, even in automated systems workers are still needed to support the automation.

- Safety monitoring means monitoring the operations beyond the conventional safety measures such as protective shields utilized by workers, safety belts, safety apron or safety shoes.

- Safety monitoring in an automated system involves the use or sensors to track the system's operation and identify conditions and events that are unsafe or potentially dangerous. The safety monitoring system is programmed to respond when it identifies unsafe conditions.

- Possible responses to various hazards might include one or more of the following:
    - Limit switches to detect proper positioning of a part in a work holding device so that the processing cycle can begin.
    - Photoelectric sensors triggered by the interruption of a light beam. This could be used to indicate that a part is arranged in proper position or to detect the presence of a human intruder into the work cell.
    - Temperature sensors to indicate that a metal work part is hot enough to proceed with a hot forging operation.
    - Heat or smoke detectors to sense fire hazards.
    - Pressure sensitive floor pads to detect human intruders into the work cell.
    - Machine vision systems to supervise the automated system and its surroundings.
    - Complete stoppage of the automated system in case of emergency.
    - Sounding an alarm when any danger is sensed.
    - Reducing the operating speed of the process in case of fault or failure.
    - Taking corrective actions to recover from the safety violation.

5.12.2 Maintenance and Repair Diagnostics

- Repair and maintenance is the most important part of any machine to keep it running successfully. Modem automated production systems are more complex and so is the problem of maintaining and repairing them.

- Maintenance and repair diagnostics refers to the capabilities of an automated system to assist in the identification of the source of malfunctions and failures of the system.

- Two modes of operation are used on a modern maintenance and repair diagnostics system.
    - **(i) Status monitoring:** In this mode, the diagnostic subsystem monitor and records the status of key sensors and parameters of the system during normal operation. When required this diagnostics system can display any of these values and provide an interpretation of current system status.
    - **(ii) Failure diagnostics:** The failure diagnostics mode is invoked when a malfunction or failure occurs. Its purpose is to interpret the current values of the monitored variables and to analyze the recorded values preceding the failure so that the cause of the failure can be identified.

Recommendation of Repair Procedure:

- This is the third mode of operation, wherein the system provides a recommended procedure to the repair crew. Methods for developing the recommendations are sometimes based on the use of expert systems in which the collective judgments of many repair experts are combined and incorporated into a computer program that uses artificial intelligence techniques.

- Status monitoring serves two important functions in machine diagnostics:
  - Providing information for diagnosing a current failure
  - Providing data to predict a future malfunction or failure.
- The first function is helpful when failure occurs in initial life of product. Initially, when a failure of the equipment has occurred, it is usually difficult for the repair crew to determine the reason for the failure and what steps should be taken to make repairs. The computer is programmed to monitor and record the variables and to draw logical inferences from their values about the reason for the malfunction. This diagnosis helps the repair personnel make the necessary repairs and replace the appropriate components.
- The second function of status monitoring is to identify signs of an impending failure, so that the affected components can be replaced before failure actually causes the system to go down. These part replacements can be made during the night shift or other time when the process is not operating, with the result that the system experiences regular operation.

5.12.3 Error Detection and Recovery

- In the operation of any automated system, there are hardware malfunctions and unexpected events that occur during operation. These events can result in costly delays and loss of production until the problem has been corrected and regular operation is restored. Traditionally, equipment malfunctions are corrected by human workers with the help of maintenance and repair diagnostics. With the increased use of computer control for manufacturing processes, there is a trend toward using the control computer not only to diagnose the malfunctions but also to automatically take the necessary corrective action to restore the system to normal operation. The term error detection and recovery is used when the computer performs these functions.
- The error detection step uses information from the system's sensor to determine when a malfunction has occurred, correctly interpret the sensor signal(s) and classify the error.
- In analyzing a given production operation, the possible errors can be classified into one of three categories of random errors, systematic errors or aberrations.
  - Random errors occur as a result of the normal stochastic nature of the process. These errors occur when the process is in statistical control. Large variations in part dimensions can cause problems in downstream operations. By detecting these deviations on a part by part basis, corrective action can be taken in subsequent operations,
  - Systematic errors are those that result from some assignable cause such as a change in raw material properties or a drift in an equipment setting. These errors usually cause the product to deviate from specifications so as to be unacceptable in quality terms.
  - Finally aberrations results from either an equipment failure or a human mistake.
- Examples of equipment failures include fracture of a mechanical shear pin, bursts in a hydraulic line, rupture of a pressure vessel and sudden failure of a cutting tool.
- Examples of human mistakes include errors in the control program, improper fixture setups and substitution of the wrong raw materials.
- The two main design problems in error detection are: (1) to anticipate all of the possible errors that can occur in a given process and (2) to specify the appropriate sensor systems and associated interpretive software so that the system is capable of recognizing each error. Solving the first problem requires a systematic evaluation of the possibilities under each of the three error classifications. If the error has not been anticipated, then the error detection subsystem cannot correctly detect and identify it.

5.13 FLEXIBLE MANUFACTURING SYSTEM

- FMS is called flexible due to the reason that it is capable of processing a variety of different part styles simultaneously at the workstation and quantities of production can be adjusted in response to changing demand patterns.
- In general, any FMS would have three main systems, (a) the GPM in the form of automated CNC machines, (b) connected by a material handling system to optimize parts flow and (c) the central control computer which controls material movements and machine flow.
- The main advantages of an FMS is its high flexibility in managing manufacturing resources like time and effort in order to manufacture a new product. The best application of an FMS is found in the production of small sets of products like those from a mass production.
- Each FMS is designed for a specific application, that is, a specific family of parts and processes. Therefore, each FMS is custom engineered; each FMS is unique.
- The use of FMS facilitates the user to find a great variety of system designs to satisfy a wide variety of application requirements.

5.14 FMS DEFINITION

- A Flexible Manufacturing System (FMS) is a production system consisting of a set of identical and/or complementary numerically controlled machine which are connected through an automated transportation system.
- Each process in FMS is controlled by a dedicated computer (FMS cell computer).
- A flexible manufacturing system (FMS) is an arrangement of machines ... interconnected by a transport system. The transporter carries work to the machines on pallets or other interface units so that work-machine registration is accurate, rapid and automatic. A central computer controls both machines and transport system.
- "FMS consists of a group of processing work stations interconnected by means of an automated material handling and storage system and controlled by integrated computer control system."

5.15 NEED OF FMS

- High productivity for all batch sizes, large or small.
- Shorter throughput times.
- Lower storage costs.
- Reduced labour if not altogether avoiding labour.
- Reduced handling.
- Flexible production system to incorporate product changes at short notice to meet customer's specific requirements.
- Sensing and taking care of such eventualities like tool breakage.

5.16 BENEFITS OF FMS

- The following benefits can be expected from an FMS:
  - There will be more utilization of machines.
  - More utilization allows less number of machines.
  - Less machines reduce floor space required.
  - There will be greater responsiveness to change.
  - Overall inventory requirements will be reduced.
  - FMS ensures lower manufacturing lead times.
  - FMS reduces direct labor requirements.
  - Allows for unattended production system, thereby reducing the number of workers.

5.17 SCOPE OF FMS

- FMS system have proved to be practical and economical for applications with the following characteristics:
  - Families of parts with similar geometric features that require similar types of equipment and processes.
  - A moderate number of tools and process steps.
  - Low to medium quantities of parts.
  - Moderate precision requirements.
- The concept of FMS was initially developed for machining applications but subsequently it was used in a variety of other manufacturing applications, such as:
  - Assembly of equipments.
  - Textile machinery manufacture.
  - Plastic injection moulding.
  - Semiconductor component manufacturing.
  - Sheet metal fabrication.
  - Welding.

5.18 TYPES OF FMS

- FMS can be classified according to
  - (i) Type of flexibility.
  - (ii) Kind of operations performed.
  - (iii) Number of machines.
  - (iv) Level of flexibility.

5.18.1 Type of Flexibility

- **Machine flexibility:** It is the capability of FMS to adapt a given machine in the system to a wide range of production operations. The greater the range of operations and part styles, the greater the machine flexibility.
- **Production flexibility:** It refers to the range of part that can be produced on the system.
- **Mix flexibility:** It is the ability to change the product mix while maintaining the same total production quantity.
- **Product flexibility:** It is the ease with which design changes can be accommodated.
- **Routing flexibility:** It is the capacity to produce parts through alternative workstation sequences in response to equipment breakdowns, tool failures, and other interruptions at individual stations.
- **Volume flexibility:** Ability to economically produce parts in high and low total quantities of production.
- **Expansion flexibility:** Ease with which the system can be expanded to increase total production quantities.

5.18.2 According to the Kinds of Operations They Perform

- Processing operations.
- Assembly operations.
- An FMS is usually designed to perform one or the other but rarely both. A difference that is applicable to machining systems is whether the system will process rotational parts or non-rotational parts. Flexible machining systems with multiple stations that process rotational parts are much less common than systems that process non-rotational parts.

5.18.3 Number of Machines

- Flexible manufacturing systems can be distinguished according to the number of machines in the system. The following are typical categories:
  - **Single machine cell (SMC):** A single machine cell consists of one CNC machining center combined with a parts storage system for unattended operation. Completed parts are periodically unloaded from the parts storage unit, and raw workparts are loaded into it.
  - **Flexible manufacturing cell (FMC):** A flexible manufacturing consists of two or three processing workstations (typically CNC machining centers or turning centers) plus a part handling system. The part handling system is connected to a load/unload station. In addition, the handling system usually includes a limited parts storage capacity.
  - **Flexible manufacturing system (FMS):** A FMS has four or more processing workstations connected by a common material handling system and a computer system. FMS generally includes non-processing workstations that support production but do not directly participate in it.

    E.g.: part/pallet washing stations, co-ordinate measuring machines, and so on.
- A FMS also consists of some additional functions like diagnostics and tool monitoring. These additional functions are needed more in an FMS than in an FMC because the FMS is more complex.

5.18.4 Level of Flexibility

- This classification can be applied to:
  - **Dedicated FMS:** A dedicated FMS is designed to produce a limited variety of part styles, and the complete universe of parts to be made on the system is known in advance. The part family is likely to be based on product commonality rather than geometric similarity. The product design is considered stable, and so the system can be designed with a certain amount of process specialization to make the operations more efficient.
  - **Random-order FMS:** Random-order FMS is more appropriate when the part family is large, there are substantial variations in part configurations, there will be new part designs introduced into the system and engineering changes in parts currently produced, and the production schedule is subject to change from day-to-day. To accommodate these variations, the random-order FMS must be more flexible than the dedicated FMS. It is equipped with general-purpose machines to deal with the variations in product and is capable of processing parts in various sequences (random-order). The dedicated FMS is less flexible but more capable of higher production rates. The random-order FMS is more flexible but at the price of lower production rates.

5.19 SUBSYSTEMS OF FMS

- There are three major subsystems in FMS:
  (i)   Computer-controlled manufacturing equipment (for example, numerically controlled machine tools, robots, gantry loaders, palletizing systems, washing stations, tool pre-setters, in-process inspection systems etc.)
  (ii) Automated materials storage, retrieval, transport and transfer system.
  (iii) Manufacturing control system (including both machine tool, tool and logistics control).
- Some FMS's have additional subsystems. For example, in a machining application there may also be systems for presetting tools, storing and retrieving tools, disposing of chips and cutting fluids, washing and inspection workpieces. These subsystems must be linked together to achieve integrated manufacturing operation.

5.20 FMS ELEMENTS/EQUIPMENTS/COMPONENTS

- There are various components of an FMS. However, the basic components are described below:
  1. Workstations,
  2. Material handling and storage system,
  3. Computer control system.

5.20.1 Workstations

- The type of work to be done decides the processing or assembly equipment used in an FMS. Most of the FMS consists of CNC machines as the major processing tool which can perform a wide variety of machining operations. However, the FMS concept is also applicable to other processes like fabrication, assembly or painting.
- Following are the types of workstations typically found in an FMS.

  (a) Load/Unload Stations: The load/unload station is the connectivity between the FMS and the rest of the factory. Raw material enters the system at this station and finished parts also exit the system from here. Loading and unloading can be done manually, although automated handling systems including robots can be utilized for it. The load/unload station should be designed to permit movement of work parts, cranes and other handling devices. The load/unload station generally include a unit to record and monitor the material flow.

  (b) Machining Stations: The most common applications of FMSs are machining operations. The workstations used in these systems are therefore predominantly CNC machine tools. Most common is the CNC machining center which possess features that make them compatible with the FMS, including automatic tool changing and tool storage, use of palletized work parts. Machining centers are generally used for non-rotational parts, whereas for cylindrical parts turning centers are used. In some machining systems, the operations are categorized as milling or turning. For milling, special milling machine modules can be used to achieve production levels higher than the capability of a machining center. For turning operations special turning modules can be designed for the FMS.

  (c) Other Processing Stations: The FMS concept has been applied to other processing operations in addition to machining. One such application is sheet metal fabrication process. The processing workstations consist of press working operations. Other operations like forging also comes under FMS. Inspection can also be incorporated into FMS. Inspection has been found to be particularly important in flexible assembly systems to ensure that components have been properly added at the workstations.

  (d) Assembly: Some FMSs are designed to perform assembly operations. Flexible automated assembly systems are being developed to replace manual labor in the assembly of products typically made in batches. Industrial robots are often used as the automated workstations in these flexible assembly systems. They can be programmed to perform tasks with variations in sequence and motion pattern to accommodate the different product styles assembled in the system.

5.20.2 Material Handling and Storage System

- The second major component of an FMS is its material handling and storage system. The Material Handling Equipment typically consists of a (i) primary handling system and (ii) a secondary handling system.
- The primary handling system is responsible for moving work parts between stations in the system. The types of primary material handling equipment typically utilized for FMS layouts are conveyors, in-line transfer system, industrial robots, AGVs etc.

- The secondary handling system consists of transfer devices, automatic pallet changers and similar mechanisms located at the workstations in the FMS. The function of the secondary handling system is to transfer work from the primary system to the machine tool or other processing station and to position the parts with sufficient accuracy and repeatability to perform the processing or assembly operation.

5.20.3 Computer Control System

- A typical FMS computer system consists of a central computer and microcomputers controlling the individual machines and other components. The central computer co-ordinates the activities of the components to achieve smooth overall operation of the system. The control system is used for work station, DNC, traffic system, pallet movement, loading or unloading of components, material handling system, tool monitoring, tool replacement, health diagnostic, data management etc.

5.20.4 Human Resources

- One additional component in the FMS is human labor. Humans are needed to manage the operations of the FMS. Functions typically performed by humans include loading raw workparts into the system, unloading finished parts from the system, changing and setting tools, equipment maintenance, repair, NC part programming and overall management of the system.

5.21 APPLICATIONS OF FMS

- FMS applications are most widely applied in machining operations.
- Other applications include:
  - Sheet metal press working,
  - Forging,
  - Assembly,
  - Surface treatment,
  - Inspection,
  - Welding.

5.22 FMS COMPARED TO OTHER TYPES OF MANUFACTURING APPROACHES

- The various manufacturing approaches can be divided into category like; conventional machines (GPM), computerized controlled conventional machines (NC or CNC), group of conventional and computerized machines (FMC), earlier group with integrated material handling system (FMS), special purpose machines (SPM) and transfer lines (TL).
- The medium and small scale industry has less orders of job, and their job keep on changing. Thus, they need not use any particular set-up. These industries works with traditional machines like the conventional lathe, milling, shaping and so on. So on a general purpose machine, low volume and variable jobs can be economically produced. It can also be a batch production where the batches repeat after certain interval of time.
- Mass production of parts needs to produce parts with repeatability and accuracy. In order to achieve this, hard wired NC and CNC machines are used. The CNC machines can run at higher speed and are programmable. These type of GPMs provides the greatest degree of flexibility.
- A large portion of the manufacturing industry involves the intermediate level of batch operations that lend themselves to the FMS approach. In this case, volume is less but varieties are more.
- FMS thus basically attempts to efficiently automate batch manufacturing operations. They are an alternative that fits in between the manual job shop and hard automation. FMS is best suited for applications that involve an intermediate level of flexibility and low or medium quantities.

- A transfer line is also called as automated flow line, is a combination of material processing unit and material handling. It consists of several individual machine tools arranged in a sequence and connected together by means of suitable material handling units. It is justified if the design of parts remain stable for sufficiently longer period and the volume of production is large.

- Fig. 5.3 shows the different types of production systems and it can be seen from the figure that FMS fits into the intermediate range of production.

- General purpose machines can accommodate a large variety of parts. They are manually operated and therefore production volumes are low. CNC machines can accommodate variety but the production volume is less as the machines are not optimized for the highest productivity for a specified type of job. It can be seen that FMC and FMS satisfy both variety and volume equally well. If we take special purpose machines, variety is much restricted. Transfer lines are dedicated usually to manufacture a component and hence can be said to have the minimum variety.

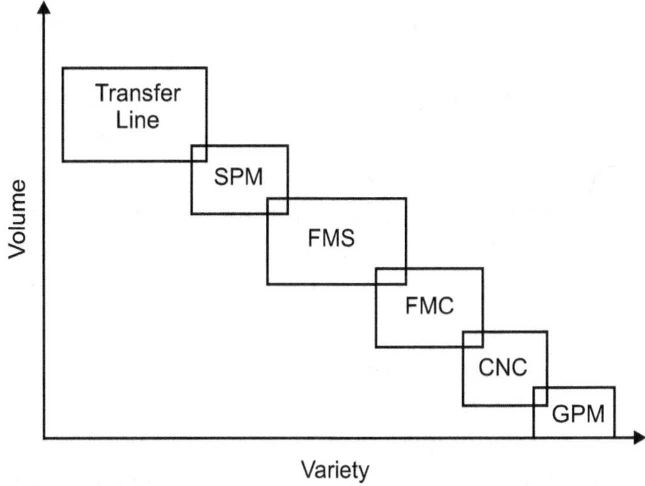

Fig. 5.3: Types of Production System

5.23 CIM

- CIM involves a fundamental strategy of integrating manufacturing facilities and systems in an enterprise through the computer and its peripheral.

- CIM can be defined in different ways depending upon its application.

- CIM involves integration of advanced technologies in various functional units of an enterprise, in an effective manner to achieve the success of the manufacturing industries.

- A deep knowledge and understanding of all the technology is required for an effective integration.

- At first, integration of advanced manufacturing technology (AMT) is required to get success in the application of CIM.

- Computers act as a subordinate to the technologies.

- Computers help, organize, and restore information in order to achieve high accuracy and speed.

- Their basic aim is to achieve the goals of the objectives within limited available capital.

- Traditionally, all the efforts were focused on achieving single goal to improve the effectiveness and competitiveness of the organization.

- But they failed because they did not satisfy the overall objectives of the manufacturing companies.

- So, a multiple goal selection was proposed to make the CIM an effective tool to improve the economy of the company.

- This new approach should improve the existing multi-criteria optimization mechanism, so that CIM can be realized globally.

- In addition, global integration approach should be applied to make globally distributed company as a single entity.

- This concept is applied to make virtual CIM more effective and hence helps in meeting the present global economic circumstances using intelligent manufacturing.

- Therefore, manufacturing technology should be blended with intelligence.

- This will help manufacturing enterprise to produce better quality.
- It will also facilitate the manufacturing equipment to solve problems posed during normal course of the operations.
- CIM basically involves the integration of advanced technologies such as CAD, CAM, CNC, robots, AMHS.
- Today CIM has moved a step ahead by including and integrating the business improvement activities such as customer satisfaction, total quality and continuous improvement.
- These activities are now managed by computers.
- Business and marketing teams continuously give the customer feedback to the design and production teams by using the networking systems.
- Based on the customer requirements, design and manufacturing teams can immediately improve the existing product design or can develop an entirely new product.
- Thus, the use of computers and automation technologies made the manufacturing industry capable to provide rapid response to the changing needs of customers.
- CIM encompasses the entire range of product development and manufacturing activities.
- And all the functions are being carried out with the help of dedicated software packages.
- The data required for various functions are passed from one application software to another in a seamless manner.
- CIM encompasses the entire range of product development and manufacturing activities.
- And all the functions are being carried out with the help of dedicated software packages.
- The data required for various functions are passed from one application software to another in a seamless manner.
- CIM reduces the human component of manufacturing and thereby relieves the process of its slow, expensive and error-prone component.
- CIM stands for a holistic and methodological approach to the activities of the manufacturing enterprise in order to achieve vast improvement in its performance.
- This methodological approach is applied to all activities from the design of the product to customer support in an integrated way, using various methods, means and techniques in order to achieve production improvement, cost reduction, fulfilment of scheduled delivery dates, quality improvement and total flexibility in the manufacturing system.
- CIM requires all those associated with a company to involve totally in the process of product development and manufacture.
- CIM also encompasses the whole lot of enabling technologies including total quality management, business process reengineering, concurrent engineering, workflow automation, enterprise resource planning and flexible manufacturing.
- A distinct feature of manufacturing today is mass customization.
- This implies that though the products are manufactured in large quantities, products must incorporate customer-specific changes to satisfy the diverse requirements of the customers.
- This requires extremely high flexibility in the manufacturing system.
- The challenge before the manufacturing engineers is illustrated in Fig. 5.4.
- Manufacturing industries strive to reduce the cost of the product continuously to remain competitive in the face of global competition.
- In addition, there is the need to improve the quality and performance levels on a continuing basis.

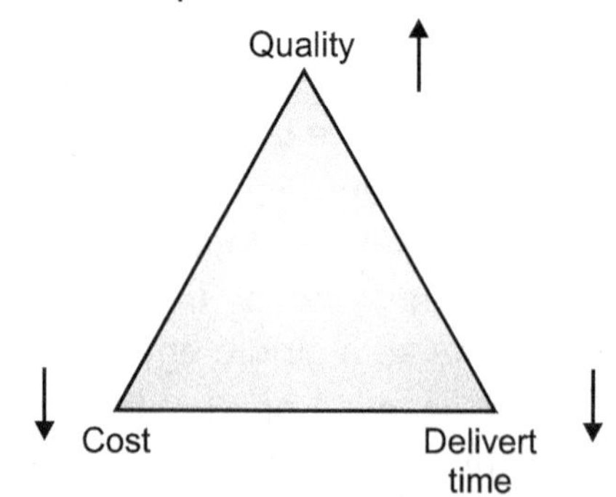

Fig. 5.3: Challenges in Manufacturing

- Another important requirement is on time delivery.
- In the context of global outsourcing and long supply chains cutting across several international borders, the task of continuously reducing delivery times is really a difficult task.
- CIM has several software tools to address the above needs.
- Manufacturing engineers are required to achieve the following objectives to be competitive in a global context.
  o Reduction in inventory
  o Lower the cost of the product
  o Reduce waste
  o Improve quality
  o Increase flexibility in manufacturing to achieve immediate and rapid response to:
    (i)   Product changes
    (ii) Production changes
    (iii) Process change
    (iv) Equipment change
    (v)   Change of personnel
- CIM technology is an enabling technology to meet the above challenges to the manufacturing.

Important Points

Automation:

- Process in which all the operations like material handling, material processing and inspection are performed automatically.
- Higher degree of mechanization where operator is a supervisor.
- Automation is the creation and application of technology to monitor and control the production and delivery of products and services.
- E.g.: Automobile assembly line, automated sugar factory, printing of newspaper.

Need of Automation:

- Reduce or eliminate routine manual and clerical tasks.
- Improve worker safety.
- Improve product quality.
- Reduce manufacturing lead time and cost of product.
- Accomplish processes that cannot be done manually.

Basic Elements of Automation:

- An automated system is used to operate some process. There may be various elements in any automated system.
- But the basic elements of an automated system are:
  o Power - to accomplish the process and operate the automated system.
  o Program of instructions – to direct the process.
  o Control system – to actuate the instructions.

Low Cost Automation:

- LCA means to implement an automation solution at the lowest cost.
- LCA is defined as 'a system where man and machine collaborate and work together, to achieve a goal with less investment'.
- LCA involves converting manual operations to automatic ones by making use of the standard equipment, mechanisms and devices of old unutilized machines, mechanisms, systems which are available free or at very lower cost.

Examples of Automation:

(a) Self driven vehicles.

(b) Machine automation in various industries.

(c) Use of robots in transfer of materials in hospitals, cleaning of passages.

(d) Automatic storage and retrieval system.

(e) Home appliances like fully automatic washing machine, toaster, coffee makers etc.

(f) Office automation.

Types of Automation:

(i) Fixed automation.

(ii) Programmable automation.

(iii) Flexible automation.

(iv) Integrated automation.

Fixed Automation:

- It is use of special purpose equipment to automate an assembly line.
- It is inflexible and cannot accommodate the variety of product.
- The product design is constant for a long period of time.
- Higher initial set-up cost but justified by mass production.
- For example, Transfer lines, automated assembly lines etc.

Programmable Automation:

- It includes equipment to accommodate a variety of product configurations.
- The sequence of operation can be modified.
- Flexible to deal with variations and changes in product configuration.
- For example, NC machine tool, PLC, industrial robots etc.

Flexible Automation:

- Capability of producing a variety of parts with virtually no time lost for changeovers from one part style to the next.
- Continuous type of production.
- Production rate is medium.
- Flexible to deal with product design variations.
- For example, Flexible manufacturing system etc.

Integrated Automation:

- Manufacturing functions are linked through an integrated computer network.
- Allow changes in product design to reduce costs and to optimize production requirements.
- For example, Computer integrated manufacturing (CIM).

Advanced Automation Function:

- Advanced automation functions include the following:
  - Safety monitoring.
  - Maintenance and repair diagnostics.
  - Error detection and recovery.

Flexible Manufacturing System:

- FMS is called flexible due to the reason that it is capable of processing a variety of different part styles simultaneously at the workstation and quantities of production can be adjusted in response to changing demand patterns.
- In general, any FMS would have three main systems, (a) the GPM in the form of automated CNC machines, (b) connected by a material handling system to optimize parts flow and (c) the central control computer which controls material movements and machine flow.
- A Flexible Manufacturing System (FMS) is a production system consisting of a set of identical and/or complementary numerically controlled machine which are connected through an automated transportation system.

Types of FMS:

- FMS can be classified according to:
    - (a) Type of flexibility.
    - (b) Kind of operations performed.
    - (c) Number of machines.
    - (d) Level of flexibility.

FMS Elements:

- There are various components of an FMS. However the basic components are described below:
    - (a) Workstations.
    - (b) Material handling and storage system.
    - (c) Computer control system.

Practice Questions

1. Define the terms Mechanization and Automation.
2. Explain the concept of automation.
3. State the need of automation.
4. What are the basic elements of automated system?
5. What is low cost automation?
6. Give some examples of automation.
7. What are the different levels of Automation?
8. What is fixed automation? State its features?
9. State the advantages of Fixed automation.
10. State the disadvantages of Fixed automation.
11. What is programmable automation? State its features.
12. State the advantages of programmable automation.
13. State the disadvantages of programmable automation.
14. What is flexible automation? State its features.
15. State the advantages of flexible automation.
16. State the disadvantages of flexible automation.
17. State the benefits of automation.
18. Describe the advanced automated function.
19. Explain the concept of FMS.
20. Define FMS.
21. State the need of FMS.
22. What are the benefits of FMS.
23. What are the scope of FMS.
24. Describe the types of FMS.
25. What are the subsystems of FMS?
26. Describe the components of FMS.
27. State the applications of FMS.
28. Compare FMS with other manufacturing processes.
29. Describe CIM.

■■■

ROBOTICS

6.1 INTRODUCTION

- The word robot was first coined by a Czech novelist Karel Capek in 1920 in a play titled Rassum's Universal Robots (RUR).
- Robot in Czech is a word for worker or servant.
- The vast majority of robots do have several qualities in common. Almost all robots have a movable body.
- Some only have motorized wheels and others have dozens of movable segments, typically made of metal or plastic. Like the bones in our body, the individual segments are connected together with joints.
- Robots spin wheels and pivot jointed segments with some sort of actuator. Various actuators can be pneumatic, hydraulic, mechanical or electric.
- A robot needs a power source to drive these actuators. Drive can be hydraulic, pneumatic, electrical or mechanical. It can also be a combination of two.
- The robot's computer controls everything attached to the circuit. To move the robot, the computer switches ON all the necessary motors and valves. Most robots are reprogrammable and their behaviour can be changed by simply changing the program.
- What many robots have in common is that they perform tasks that are too dull, dirty, delicate or dangerous for people.
- Usually, we also expect them to be autonomous, that is, to work using their own sensors and intelligence, without the constant need for a human to control them.

6.2 DEFINITION OF ROBOT AND ROBOTICS

- *Robot* can be defined as *'an automatic apparatus or device that performs functions ordinarily attributed to humans or operates with what appears to be almost human intelligence'*.
- Robotics Institute of America defines robot as, "A robot is a reprogrammable multifunctional manipulator designed to move material, parts, tools or specialized devices through variable programmed motions for the performance of a variety of tasks".
- The robot is a computer-controlled device that combines the technology of digital computers with the technology of servo control of articulated chains. It should be easily reprogrammed to perform a variety of tasks and must have sensors that enable it to react and adapt to changing conditions.
- *Robot* may be defined as *'a system that contains sensors, control systems, manipulators, power supplies and software all working together to perform a task'*.
- A mechanical device with joints and links, guided by sensors and driven by actuators, controlled by a program, that handle and manipulate parts, materials, tools and devices for performing different tasks in various work conditions can be called as a robot.
- *Robotics* is a branch of engineering that involves the conception, design, manufacture and operation of robots. This field overlaps with electronics, computer science, artificial intelligence, mechatronics, nanotechnology and bioengineering.

6.3 LAWS OF ROBOTICS

- Isaac Asimov proposed three "Laws of Robotics" in 1950 and later added the "zeroth law".

Zeroth Law:
- A robot may not injure humanity, or through inaction, allow humanity to come to harm.

First Law:
- A robot may not injure a human being, or through inaction, allow a human being to come to harm, unless this would violate a higher order law.

Second Law:
- A robot must obey the orders given to it by human beings except where such orders would conflict with the First Law.

Third Law:
- A robot must protect its own existence as long as such protection does not conflict with the First or Second Law.

6.4 CHARACTERISTICS OF ROBOT

- The essential characteristics of any Robot are explained below:

Sensing:
- Human has five senses through which he can sense his surroundings. Similarly a robot should have the ability to sense its surroundings. This needs a lot of sensors to be incorporated in the robot.
- The sensors like light sensors (eyes), touch and pressure sensors (hands), chemical sensors (nose), hearing and sonar sensors (ears) and taste sensors (tongue) will give a robot awareness of its environment.

Movement:
- A robot needs to be able to move around its surrounding. Whether rolling on wheels, walking on legs or propelling by thrusters, a robot needs to be able to move.
- Either a whole robot should move or just parts of the robot should move.

Energy:
- A robot needs to be able to power itself. A robot might be solar powered, electrically powered, battery powered.
- The source of energy for a robot depends on the application for which it is designed.

Intelligence:
- Human brain is a thing which cannot be artificially developed. But to work as a human does, robot needs some kind of smartness.
- This is possible by means of programming. A program is stored in the robot controller which can be called as the *brain of robot*.
- This gives the required logic and artificial intelligence to the robot.

6.5 DEGREE OF FREEDOM

- Degrees of freedom (DOF) "a term that describes a robot's freedom of motion in three dimensional space".
- Degree of freedom for a robot is defined as "the number of independent movements performed by the robot wrist in three dimensional space, relative to the robot's base".
- It refers to the ability of the robot arm to move forward and backward, up and down and to the left and to the right.
- For each degree of freedom, a joint is required. A robot requires six degrees of freedom to be completely versatile.
- The number of degrees of freedom defines the robot's configuration.
- Many robotic applications require movement in all the three directions; i.e. X, Y and Z.
- These tasks require three joints, or three degrees of freedom.
- The three degrees of freedom in the robot arm are 1 rotational and 2 linear.
    - **(a) Rotational movement:** The rotational movement is about a vertical axis. Because of this, the robot can swivel its arm about its base.

(b) Radial movement: The linear movement in horizontal direction allows the extension and retraction of the arm relative to the base. This movement allows the robot arm to cover large area along radial direction.

(c) Vertical movement: The linear movement about the vertical axis provides vertical lift to the robot arm.

- These degree of freedom of robots arm are shown in Fig. 6.1 (a).
- For applications that require more freedom, additional degrees can be obtained from the wrist, which can be obtained from the end effector.
- These additional three degrees of freedom in the wrist provide rotary motion to the arm.
- These degree of freedom are named as: pitch, yaw and roll as shown in the Fig. 6.1 (b).

(a) Pitch: The pitch, or bend, is the up-and-down movement of the wrist about the horizontal axis.

(b) Yaw: The yaw is the side-to-side movement, or swivel about the vertical axis in horizontal plane.

(c) Roll: The roll is the rotation or swivel about the arm axis (horizontal).

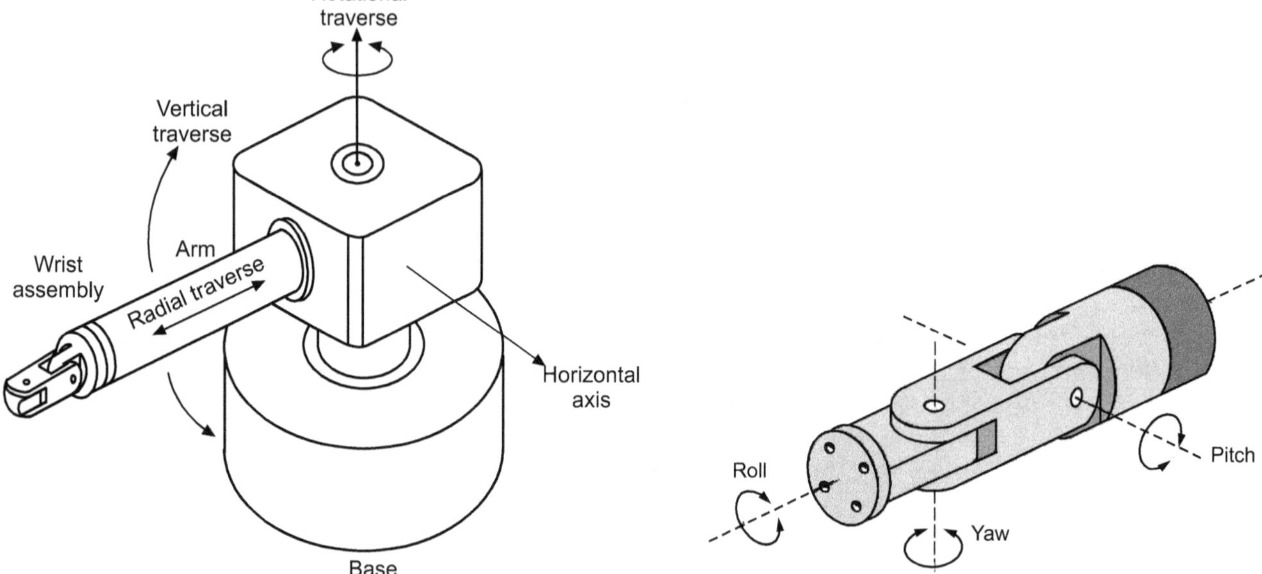

(a) Degrees of Freedom of Robot Arm   **(b) Degrees of Freedom of Robot Wrist**

Fig. 6.1

- Thus, the first three degrees of freedom are for arm movement while the remaining are for wrist movement.
- A robot requires a total of six degrees of freedom to locate and orient its hand at any point in its work envelope, Fig. 6.2.

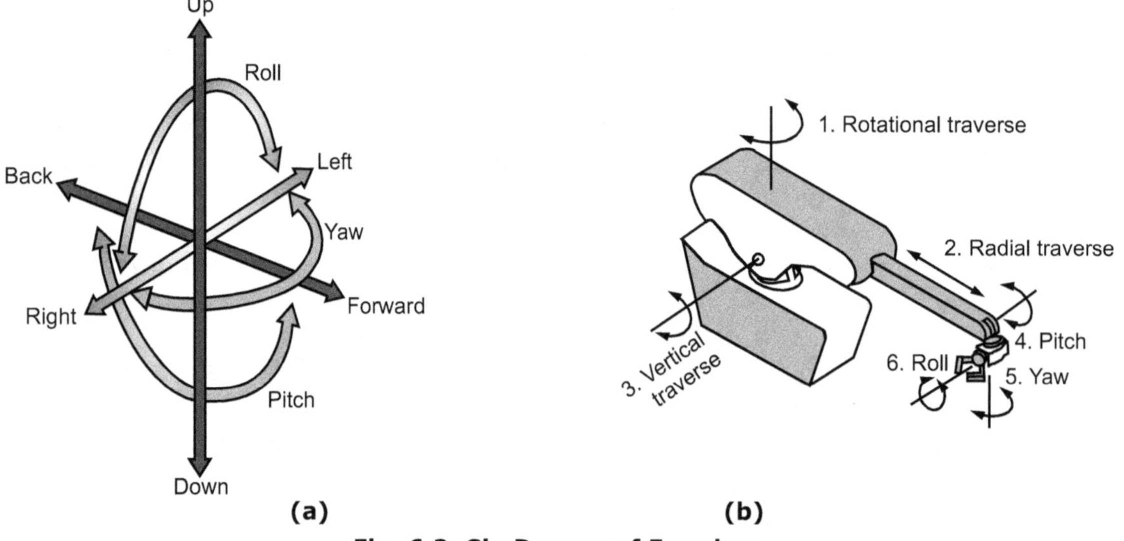

(a)   **(b)**

Fig. 6.2: Six Degree of Freedom

6.6 BASIC DEFINITIONS

Axis/Axes:
- An axis is a line across which a rotating body turns. Two axes are required to reach any point in a straight plane, while three axes (X, Y, Z) are needed to reach any point in space.
- Three further axes (roll, pitch and yaw) are needed to control the orientation of the end of the robot arm or wrist.

Number of axes:
- Two axes are required to reach any point in a plane; three axes are required to reach any point in space.
- To fully control the orientation of the end of the arm (i.e. the wrist) three more axes (yaw, pitch and roll) are required.

End effector:
- Also known as end of arm tooling, is the 'hand' attached to the end of the robot arm or wrist.
- End effectors include grippers, vacuum cups, spray guns, welding tools and electro-magnetic pick-ups, their performance being vital to precision and repeatability.

Working envelope:
- It is also known as *work space* or *work volume*. It is the region of space the end effector of the manipulator can reach.

Actuators:
- These are devices that convert electrical, hydraulic and pneumatic energy into robot motion.

Payload:
- It is the maximum load or carrying capacity, including weight of the end effector.

Reach:
- It is the maximum distance a robot can extend its arm to perform a task.

Manipulator:
- This refers to the arm mechanism, created from a sequence of joint and linkage combinations, including the wrist.

Joints:
- A part that provides relative motion between the two parts. In robotics joints provides either sliding or rotating motion to the arm.

6.7 WORK ENVELOPE OF ROBOT

- Work envelope is also known as *work space* or *work volume*. It is the region or space the end effector of the manipulator can reach.
- The work envelope is the range of motion over which a robot arm can move.
- In practice, it is the set of points in space that the end effector can reach.
- The size and shape of the work envelope depends on the co-ordinate geometry of the robot arm and also on the number of degrees of freedom.
- Some work envelopes are flat, confined almost entirely to one horizontal plane, others are cylindrical, while some are spherical. Some work envelopes have complicated shapes.
- Fig. 6.3 shows a simple example of a work envelope for a robot arm using cylindrical co-ordinate geometry. The set of points that the end effector can reach, lies within two concentric cylinders, labelled "inner limit" and "outer limit". The work envelope for this robot arm is shaped like a new roll of wrapping tape.

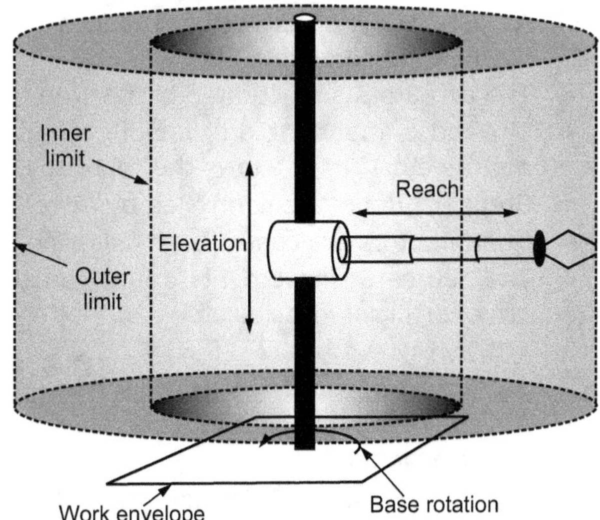

Fig. 6.3: Work Envelope of Robot

- When choosing a robot arm for a certain industrial purpose, it is important that the work envelope be large enough to encompass all the points that the robot arm will be required to reach. But it is wasteful to use a robot arm with a work envelope much bigger than necessary.

6.8 BASIC COMPONENTS OF ROBOT

- Any industrial robot has to have the following basic parts to perform given work/job satisfactorily:

6.8.1 Arm

- The arm is the part of the robot that positions the end-effector and sensors to do their pre-programmed task.
- Most of the robot arms resemble human arms and have shoulders, elbows, wrists, even fingers.
- With the movement of robot arm, the shoulder, elbow and wrist move and twist to position the end effector in the exact right spot.
- This gives the robot a lot of ways to position itself in its environment. Each joint gives the robot one degree of freedom.
- It means that a simple robot arm with three degrees of freedom could move in any of the three directions; up and down, left and right, forward and backward, rotation about base etc.
- Industrial robot arms can vary in size and shape.

6.8.2 End-effectors

- A hand of a robot is considered as end effectors. The effectors are the parts of the robot that actually do the work.
- The grippers and tools are the two significant types of end effectors that helps the robot to do its job.
- The grippers are used to pick and place an object, while the tools are used to carry out operations like spray painting, spot welding etc. on a work piece.
- If the robot has more than one arm, there can be more than one end-effector on the same robot, each suited for a specific task.
- Some robots can change end-effectors and be reprogrammed for a different set of tasks.

6.8.3 Actuators

- An actuator is a device that produces translatory or rotary movement in the links or makes the freedoms possible.
- The actuators are actuated in reaction to the feedback from the sensors.
- The most important and popular actuator is a motor (DC motors, stepper motors, servo motors etc.) which allows the robot to control a wheel, a switch or even an arm.
- Stepper motors are controlled by increasing or decreasing the voltage.
- Servo motors are controlled by slowing or speeding up the motor using a feedback loop.
- The source of movement (drive) can be electric drive (motor), hydraulic drive, pneumatic drive, artificial muscles etc.

6.8.4 Sensors

- A sensor is an element in a control system that acquires a physical parameter and changes it into a signal.
- The sensor sends information, in the form of electronic signals back to the controller. Sensors also give the robot controller information about its surroundings and let it know the exact position of the arm, or the state of the world around it.

- Sensors can also assist end effectors by adjusting for part variances.
- To know the position of each joint in the mechanical linkage, potentiometers or encoders are used as sensors.
- Tachometers or accelerometers are used to measure the velocity and/or acceleration at each joint.
- Vision sensors (cameras, laser range finders), acoustic sensors (ultrasonic ranging systems), touch sensors are used to detect objects or features in the environment.
- Microphones allow robots to detect sounds. Sensors such as buttons embedded in bumpers can allow the robot to determine when it has collided with an object or a wall.

6.8.5 Controllers

- Every robot is connected to a computer that co-ordinates both the arm to work together. This computer is known as the *controller*.
- The controller functions as the "brain" of the robot.
- The controller also allows the robot to be networked to other systems, so that it may work together with other machines, processes or robots.
- The robotic arm controller runs a set of instructions written in code called a *program*.
- The controller actually is some electronic device which receives the signals from the sensors and provides it to the drive to actuate the actuators.
- The actuation produced is fed back to the controller by the feedback element, which is compared with the given input and further action is taken.

6.8.6 Drive

- The drive is the engine or motor that moves the links into their designated positions. The links are the sections between the joints.
- Industrial robot arms generally use one of the following types of drives: mechanical, hydraulic, electric or pneumatic.
- Hydraulic drive systems give a robot great speed and strength.
- An electric system provides a robot with less speed and strength.
- Pneumatic drive systems are used for smaller robots that have fewer axes of movement.

6.8.7 Software

- The software used consists of several levels. Motor control software consists of algorithms which help the servo to move smoothly utilizing the data from feed-back units.
- At the next level, there is software to plan the trajectory of the end effector and translate the same into commands to individual motor controllers.
- The output of sensors is also to be interpreted and decisions made.
- At the highest level, there is software which accepts commands from the user of the robot and translates it into appropriate actions at the lower level.

6.9 WORKING OF ROBOT

- An industrial robot has the basic parts like arm, sensor, actuator, controller etc.
- These subsystems communicate among themselves via interfaces, whose function consists basically of decoding the transmitted information from one medium to another.
- Fig. 6.4 shows the block diagram representation of a typical robotic mechanical system.
- The input is a prescribed task, which is defined earlier.
- The output of a robotic mechanical system is the actual task, which is monitored by the sensors. These sensors sense and transmit the information in the form of feedback signals.
- This is compared with the predefined task given to the controller.
- The errors between the prescribed and the actual task are then fed back into the controller, which then synthesizes the necessary corrective signals.
- These are in turn fed back to the actuators, which then drive the mechanical system through the required task.
- Thus, the given task is performed by the robot.

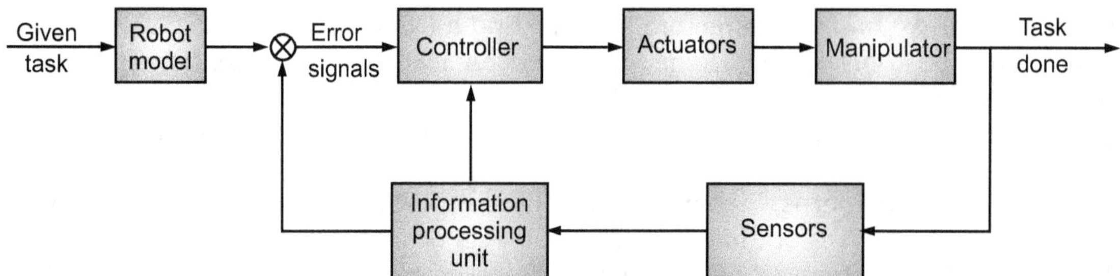

Fig. 6.4: General Structure of Robotic Mechanism

6.10 ROBOT ANATOMY

- The Anatomy of Industrial Robots deals with the assembling of outer components of a robot such as wrist, arm and body.
- Robot manipulator consists of two sections:
  - (i) Body-and-arm assembly – with typically three degrees-of-freedom for positioning of objects in the robot's work volume.
  - (ii) Wrist assembly with typically three degrees-of-freedom for orientation of objects.
- At the end of the manipulator's wrist assembly is a device related to the task that must be accomplished by the robot. The device is called an end effector and can be either a gripper for holding a work part, or a specific tool for performing some process.
- The body-and-arm of the robot is used to position the end effector and the robot's wrist is used to orient the end effector.
- In total, a robot typically has up to six degrees of freedom to position and orient the end effector to a point or line in space.

6.11 JOINTS AND LINKS

- The manipulator of an industrial robot consists of a series of joints and links. Different joints and links encompass the study of robot anatomy.
- Joints provide relative motion between two links of the robot whereas links are rigid members between joints.
- The two basic types of joint motion are linear and rotary.
- Each joint provides a "degree-of-freedom" of motion. In most cases only one degree-of-freedom is associated with each joint. So a robot's complexity can be classified according to the total number of degrees-of-freedom they possess.
- Each joint is connected to two links, an input link and an output link, with the joint providing controlled relative movement between the input link and output link.
- Most robots are mounted upon a stationary base, such as the floor. From this basis a joint-link numbering scheme may be recognised (see Fig. 6.5).

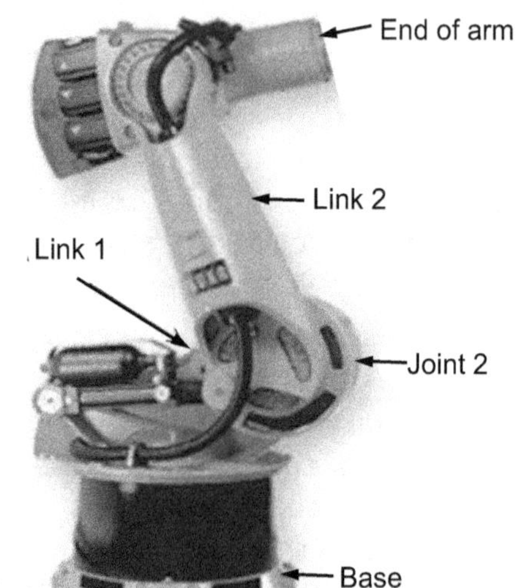

Fig. 6.5: Joint-link Numbering Scheme

- The joint-link numbering scheme counts all the joints and links which comprise the robotic system. The number scheme is ordered from the robot's base to the last joint and link in the robotic system.
- The robotic base and its connection to the first joint is termed as link 0 and the first joint in the sequence is joint 1; link 0, therefore, is the input link for joint 1, while the output link

from joint 1 is link 1 that leads to joint 2. Thus, link 1 is simultaneously, the output link for joint 1 and the input link for joint 2. This joint-link numbering scheme follows in the same fashion for all joints and links in the robotic system.

- Five types of mechanical joints for robots are explained below:

6.11.1 Linear Joint

- The linear joint provides the translational sliding motion between the input and output link.
- The axes of both the links are parallel to one another.
- Such a joint is called as 'type L' joint.
- The linear joint is shown in Fig. 6.6.

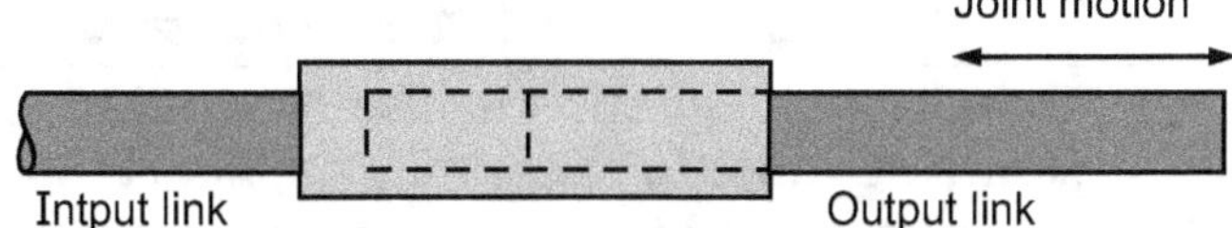

Fig. 6.6: Linear Joint

6.11.2 Orthogonal Joint

- The orthogonal joint provides the translational sliding motion between the input and output link.
- The axis of output link is perpendicular to that of the input link.
- Such a joint is called as type O joint.
- The orthogonal joint is shown in Fig. 6.7.

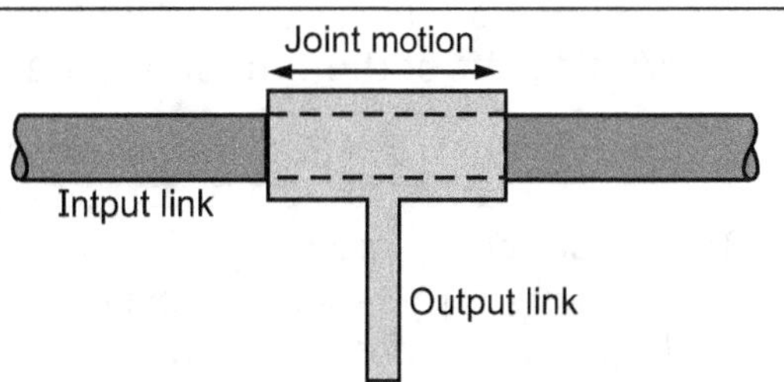

Fig. 6.7: Orthogonal Joint

6.11.3 Rotational Joint

- The rotational joint provides the relative rotational motion between the input and output link.
- The axis of rotation is perpendicular to the axes of input and output link.
- Such a joint is called as 'type R' joint.
- The rotational joint is shown in Fig. 6.8.

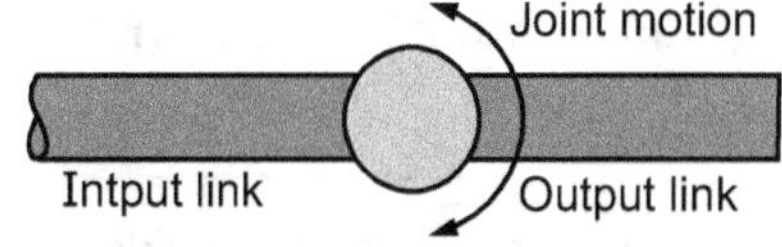

Fig. 6.8: Rotational Joint

6.11.4 Twisting Joint

- The twisting joint provides the relative rotational motion between the input and output link.
- The axis of rotation is parallel to the axes of input and output link.
- Such a joint is called as 'type T' joint.
- The twisting joint is shown in Fig. 6.9.

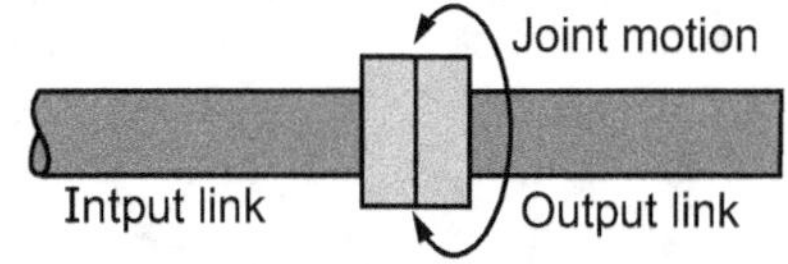

Fig. 6.9: Twisting Joint

6.11.5 Revolving Joint

- The revolving joint provides the relative rotational motion between the input and output link.
- The axis of the input link is parallel to the axis of rotation of the joint.
- The axis of the output link is perpendicular to the axis of rotation of the joint.
- Such a joint is called as 'type V' joint.
- The revolving joint is shown in Fig. 6.10.

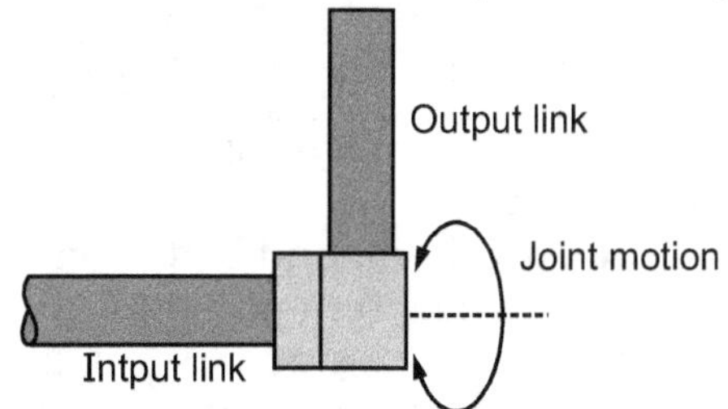

Fig. 6.10: Revolving Joint

6.12 BASIC ROBOT CONFIGURATIONS

- Industrial robots come in a variety of shape and sizes and different joint motions.
- Depending on the joints, a robot can possess number of motions, but the basic arm and body configuration is limited by the degree of freedom.
- As only three degree of freedom is associated with the arm and body, the basic robot configuration in industrial robots commonly available are explained below.
  (a) Polar configuration, (b) Cylindrical, (c) Cartesian co-ordinate, (d) Joint arm, (d) SCARA.

6.13 POLAR CONFIGURATION

- The polar configuration is also referred to as the spherical configuration as this combination allows the robot to operate in a spherical work volume.
- The polar arm and body configuration is shown in Fig. 6.11.
- In the polar configuration, the robot arm has following movements:
  1. Linear movement that allows the arm to extend and retract because of one linear joint.
  2. Rotary movement that occurs around an axis (vertical) perpendicular to the base because of one twisting joint.
  3. Vertical lift of the arm about the pivot point because of one rotational joint.

6.13.1 Advantages of Polar Configuration

- The polar configuration has following advantages:
  1. It allows long reach capability in the horizontal position.
  2. It provides good load lifting capabilities.
  3. It is suitable for applications where a small amount of vertical movement is adequate, such as loading and unloading of components.

6.13.2 Limitations of Polar Configuration

- The polar configuration has following limitations:
  1. It has a low vertical reach.
  2. It has a reduced mechanical rigidity.

6.13.3 Applications of Polar Configuration

- The polar configuration finds applications in machine tool loading, movement of material, stacking of components, heat treatment operations forging etc.

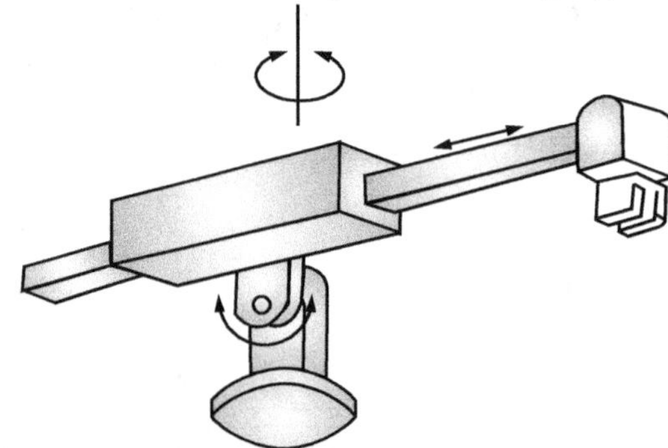

Fig. 6.11: Polar Arm and Body Configuration

6.14 CYLINDRICAL CONFIGURATION

- The cylindrical configuration allows the robot to reach the work space in a rotary movement like a cylinder.
- The cylindrical arm and body configuration is shown in Fig. 6.12.
- In the cylindrical configuration, the robot arm has following movements:
  1. Rotational movement of the column about its axis because of one twisting joint.
  2. Linear movement of the assembly along the column because of one linear joint.
  3. Linear movement in and out, relative to the column axis because of one orthogonal joint.

6.14.1 Advantages of Cylindrical Configuration

- The cylindrical configuration has following advantages:
  1. It has higher load carrying capacity.
  2. It provides high rigidity to the manipulator.
  3. It is generally suitable for pick and place applications.

6.14.2 Limitations of Cylindrical Configuration

- The cylindrical configuration has following limitations:
  1. It requires more floor space.
  2. It has a reduced mechanical rigidity because robots with a rotary axis must overcome the inertia of the object when rotating.

6.14.3 Applications of Cylindrical Configuration

- The cylindrical configuration finds applications in:
  1. Conveyor pallet transfers,
  2. Machine tool loading,
  3. Forging applications,
  4. Packing operation,
  5. Precision small assembly etc.

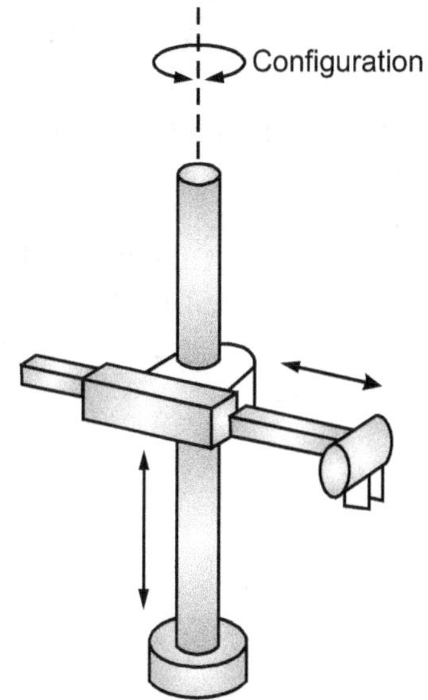

Fig. 6.12: Cylindrical Body Configuration

6.15 CARTESIAN CO-ORDINATE CONFIGURATION

- The Cartesian co-ordinate robot is also referred to as rectilinear robot or X-Y-Z robot of the spherical configuration because it is equipped with three sliding joints for assembling XYZ axes.
- The Cartesian co-ordinate robot arm and body configuration is shown in Fig. 6.13.
- The arm movement of a robot using the Cartesian configuration can be described by three intersecting perpendicular straight lines, referred to as the X, Y and Z axes.
- In the Cartesian co-ordinate configuration, the robot arm has following movements:
  1. Linear movement that allows vertical lift to the arm because of one linear joint.
  2. Two sliding movements those are perpendicular to each other because of two orthogonal joint.
- This configuration robot will process in a rectangular work space by means of this three joints movement.
- Because movement of the arm can start and stop simultaneously along all three axes, motion of the tool tip is smoother.
- This allows the robot to move directly to its designated point, instead of following trajectories parallel to each axis.

6.15.1 Advantages of Cartesian Co-ordinate Configuration

- The Cartesian co-ordinate configuration has following advantages:
  1. It has high load carrying capability.
  2. It provides a rigid structure and also a high degree of mechanical rigidity and accuracy.
  3. It can produce high repeatability with least error and at a good speed.

6.15.2 Limitations of Cartesian Co-ordinate Configuration

- The Cartesian co-ordinate robot configuration has following limitations:
  1. It has a small and rectangular work envelope.
  2. It has a reduced flexibility.

6.15.3 Applications of Cartesian Co-ordinates

- The Cartesian co-ordinate configuretion finds applications in inspection, assembly, machining operations, welding, finishing operations etc.
- This configuration also get a name of Gantry Robot as it is capable of carrying high loads with the help of its rigid structure and this weight lifting capacity does not vary at different locations within the work envelope.

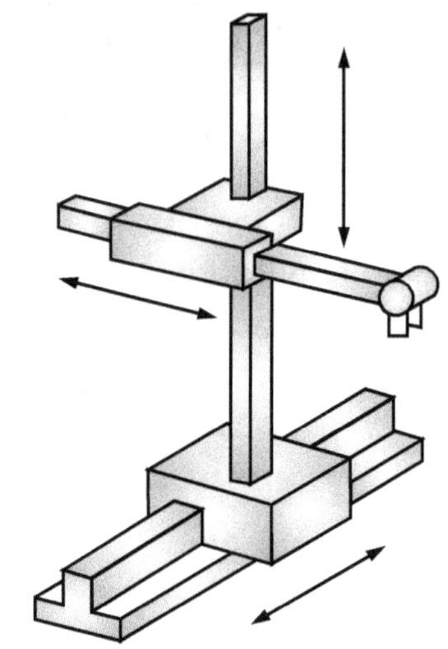

Fig. 6.13: Cartesian Robot

6.16 JOINTED ARM CONFIGURATION

- The jointed arm configuration resembles to a human arm.
- The jointed arm and body configuration is shown in Fig. 6.14.
- In this configuration, the robot arm has following movements:
  1. Rotary movement (vertical column that swivels about the base) that occurs around an axis (horizontal) parallel to the base because of one twisting joint.
  2. Rotary movement at the top of the column about the shoulder joint (along horizontal axis) because of one rotational joint.
  3. Rotary movement at the output arm about the elbow joint (along horizontal axis) because of one rotational joint.
- It gets three rotary joints and three wrist axes, which form into six degrees of freedoms. As a result, it has the capability to be controlled at any adjustments in the work space.

6.16.1 Advantages of Jointed Arm Configuration

- The jointed arm configuration has following advantages:
  1. It has a huge work volume.
  2. It has higher flexibility and so is quick in operation.
  3. Two rotational joints allows for higher reach from the base.
  4. It provides reaching the congested small openings without restrictions.

6.16.2 Limitations of Jointed Arm Configuration

- The jointed arm configuration has following limitations:
    1. It has a difficult operating procedure.
    2. It has a plenty of components.

6.16.3 Applications of Jointed Arm Configuration

- The jointed arm configuration finds applications in spray painting, spot welding, arc welding etc.
- This configuration robot is also called as *articulated robot*.

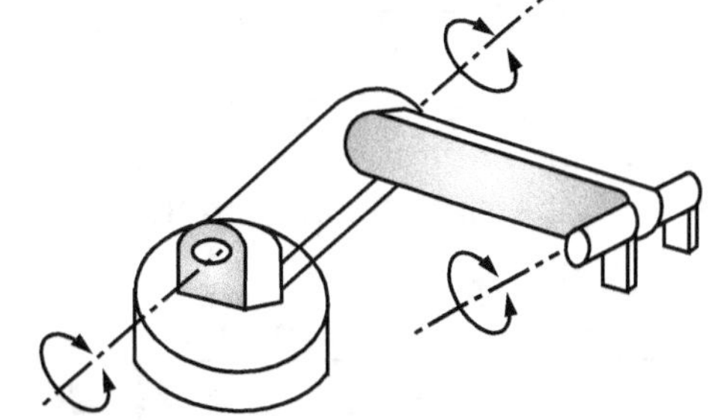

Fig. 6.14: Jointed Arm Body Configuration

6.17 SCARA

- SCARA is an acronym for Selective Compliance Assembly Robot arm.
- One Scara configuration is shown in Fig. 6.15.
- In Scara, the robot arm has following movements:
    1. Linear movement that allows the arm to extend and retract because of one orthogonal joint.
    2. Rotary movement at the top of the column about the shoulder joint (along vertical axis) because of one revolving joint.
    3. Rotary movement at the output arm about the elbow joint (along vertical axis) because of one rotational joint.

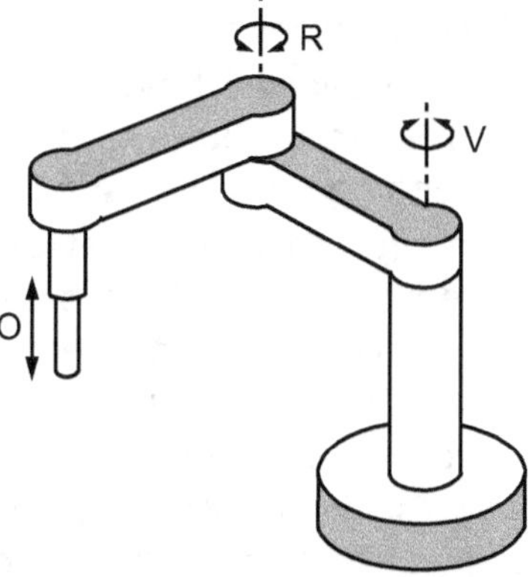

Fig. 6.15: SCARA Configuration

6.18 ROBOT SENSORS

- Sensors are devices that can sense and measure physical properties of the environment like temperature, luminance, resistance to touch, weight, size etc.
- The potential range of robotic applications requires different types of sensors to perform different kinds of sensing tasks.
- Various sensing needs includes orientation, displacement, velocity, acceleration and force.
- Robots must also sense the characteristics of the tools and materials they work with.
- There are generally two categories of sensors used in robotics. These are sensors for internal feedback and for external feedback.
- Internal sensors are used to monitor and control the position and velocity of various joints of the robot. They form a feedback control loop with the robot controller.
- Examples of internal sensors include potentiometers, resolvers or encoders, while tachometers of various types are deployed to control the speed of the robot arm.
- External sensors are external to the robot itself and are used when we wish to control the operations of the robot in its work environment.
- External sensors are simple devices, such as limit switches that determine whether a part has been positioned properly, or whether a part is ready to be picked up from an unloading bay.

6.18.1 Functions of Robot Sensors

- The sensors mounted on the robot serve the following functions:
    1. To detect the position of an object.
    2. To apply safe force to grip an object.
    3. To position an object accurately.
    4. To prevent collision of robot.
    5. To inspect the objects in production lines.

6.19 CLASSIFICATION OF SENSORS

- Though currently available sensors rely on different physical properties for their operation, they may be classified into following general types.
    1. Contact type sensor
    2. Non-contact type sensor.

6.20 CONTACT SENSOR

- The sensors that operate on touch are called as contact sensors. As these sensors must touch their environment to operate, their use is limited to objects and conditions that can do no harm to the sensors.
- These sensors are also called as touch and tactile sensors.
- Contact sensors have varying sensitivity and complexity. Some can only determine whether something is touching or not, while others accurately measure the pressure of the contact depending on the elastic limit of the sensors.
- A tactile sensor is a device that measures information arising from physical interaction with its environment. Tactile sensors are used to determine whether contact is made between the sensor and another object. Tactile sensors can be divided into two types in robot applications: (1) touch sensors and (2) force sensors,
- Touch sensors are those that indicate simply that contact has been made with the object. The touch sensors are basically used to determine whether the gripper jaw touches any object. This may also send a signal to the controller that target is reached, collision may occur. A micro-switch is another example of this type of sensor. It sends a signal to the controller when a contact is made. A touch sensor is generally a single-point contact, though the sensory area can be of any size.
- Force sensors are used to indicate the magnitude of the force with the object. This might be useful in a gripper to measure and control the force being applied to grasp an object.
- They are divided into following types:
    o   Force/ torque sensor.
    o   Dynamic sensor.
    o   Thermal sensor.
    o   Capacitive sensors.
    o   Piezoresistive sensors.
    o   Magnetic sensors.

6.20.1 Touch Sensors

- Thermal sensors are used to identify the materials of the objects made.
- Thermal sensing involves detecting thermal gradients in the skin, which are correspondent to both the temperature and the thermal conductivity of an object.

6.20.2 Force Sensors

- Force sensors can measure force between almost any two surfaces and are durable enough to stand up to most environments.

- It is used to indicate the magnitude of force with the object.

- These sensors are located between the gripper and the last joint of the wrist, so that the forces can be measured.

- Force measurement is useful in grasping delicate objects by control of force.

- A load cell is a type of transducer, specifically a force transducer. It converts a force such as tension, compression, pressure, or torque into an electrical signal that can be measured and standardized. As the force applied to the load cell increases, the electrical signal changes proportionally.

6.20.3 Capacitive Sensors

- These sensors operate by measuring the variations of capacitance from an applied load over a parallel plate capacitor.

- The capacitance is related to the separation and area of the parallel plate capacitor, which uses an elastomeric separator to provide compliance.

- Capacitive sensors can be fabricated in very small sizes.

6.20.4 Linear Variable Differential Transformer

- This sensor is used to measure linear displacement and works on the same principle as the AC transformer that is used to measure movement.

- It is a positional sensor that does not suffer from mechanical wear problem.

- It basically consists of three coils, one primary and two secondary and an armature (core). The secondary coils are connected electrically together in series but 180° out of phase on either side of the primary coil.

- The armature is basically a ferromagnetic core such as iron and is connected to the object being measured and slides or moves up and down inside the tubular body of the LVDT.

- An AC voltage is applied to the primary coil which in turn induces an EMF signal into the two adjacent secondary coils.

- If the soft iron magnetic core armature is exactly in the centre of the windings, both the secondary coils produces equal and opposite voltage. Hence, the output voltage (V_{out}) is zero.

- When the core moves to the left or right direction, the output voltage of one secondary coil decreases and the output voltage of the other secondary coil increases.

- The polarity of the output signal depends upon the direction and displacement of the moving core. The greater the movement of the soft iron core from its central null position the greater will be the resulting output signal. This output voltage is used to determine the displacement.

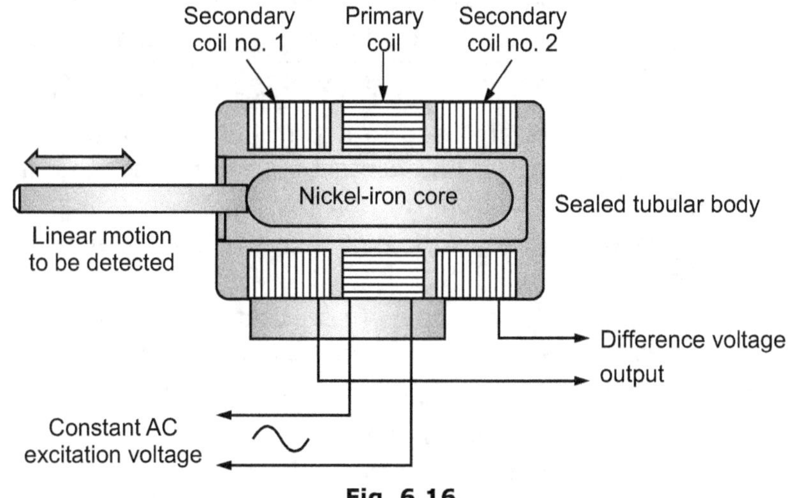

Fig. 6.16

- The result is a differential voltage output which varies linearly with the cores position. Therefore, the output signal has both an amplitude that is a linear function of the cores displacement and a polarity that indicates direction of movement.

6.20.5 Potentiometer

- A potentiometer is basically a three-terminal resistor with a sliding or rotating contact that is used to vary resistance in a circuit.
- A potentiometer is a resistive sensor used to measure linear displacements as well as rotary motion.
- They have three terminals, where the one in the middle is known as the *wiper* and the other two are known as *ends*. The wiper is a movable contact where resistance is measured with respect to it and either one of the end terminals.
- The wiper travels along the strip when the knob is turned. The closer the wiper is to the end terminal, to which it is wired, the less is the resistance, because the path of the current will be shorter. The further away it moves from the terminal, the greater the resistance will be.
- The output voltage is proportional to the distance travelled. There are two types of potentiometer, linear and rotary potentiometer. The linear potentiometer has a slide or wiper. The rotary potentiometer can be a single turn or multi turn.

6.21 NON-CONTACTING TYPE SENSORS

- The sensors those gather information without touching an object are called as *non-contact sensors*.
- They can be used in environments where contact sensors would get damaged or wear out quickly.
- These sensors can sense most materials, including liquid, powder and smoke, and can also measure many parameters, including velocity, position and orientation.
- Simple non-contact sensors merely determine whether something is present or not more complicated devices can be used to distinguish between objects and work piece.
- The non-contact sensors are classified into following types:

   (i) Optical type sensors and (ii) Proximity sensors.

6.21.1 Optical Type Sensors

- Photocells and other photometric devices that are used to detect the presence or absence of objects. Often used in conjunction with proximity sensors.

6.22 PROXIMITY SENSORS

- These sensors are used to determine that an object is close to the sensor before it makes contact. These are non-contact type of devices that indicate the presence of an object within specific distance from the sensor.
- These sensors send the signal to the controller which alerts the gripper about presence of object.
- It works on the principle of reflected light waves, elapsed time for reflected sound, intensity induced eddy currents, magnetic fields, back pressure from air jets etc.
- The different types of proximity sensors are:
   1. Magnetic proximity sensor
   2. Optical proximity sensor
   3. Ultrasonic proximity sensor
   4. Inductive proximity sensor
   5. Capacitive proximity sensor
   6. Eddy current proximity sensor

6.22.1 Ultrasonic Proximity Sensors

- An ultrasonic proximity sensor uses a piezoelectric transducer to send and detect sound waves.

- Transducer generate high frequency sound waves and evaluate the echo by the detector which is received back after reflecting OFF the target.

- Sensors calculate the time interval between sending the signal and receiving the echo to determine the distance to the target. When the target enters the operating range the output switches.

- The ultrasonic proximity switches are equipped with temperature sensors and a compensation circuit, in order to be able to compensate for changes in operating distance caused by temperature fluctuations.

- The advantages of an ultrasonic proximity sensor are, no physical contact with the object to be detected, therefore, no friction and wear, not affected by target appearance or environment around it, can work in adverse conditions.

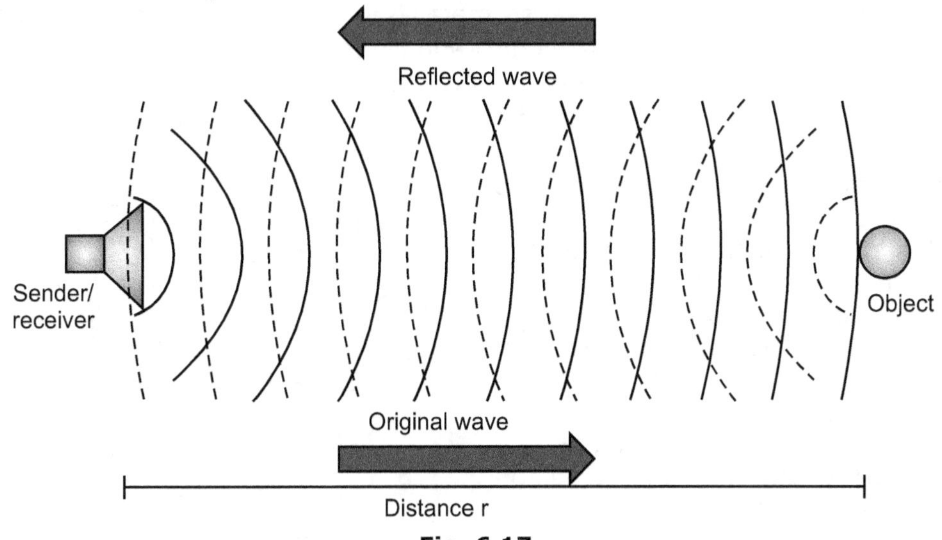

Fig. 6.17

6.22.2 Inductive Proximity Sensor

- Inductive proximity sensors operate under the electrical principle of inductance. Inductance is the phenomenon where a fluctuating current, which by definition has a magnetic component, induces an electromotive force (emf) in a target object. To amplify a device's inductance effect, a sensor manufacturer twists wire into a tight coil and runs a current through it.

- An inductive proximity sensor has four components; the coil, oscillator, detection circuit and output circuit.

- The coil generates the high frequency magnetic field in front of the face. When the metallic target comes in this magnetic field, Eddy circuits build up in the metallic object.

- This reduces the Inductive sensor's own oscillation field. This is detected by the sensor's detector circuit.

- If the oscillation amplitude reaches a certain threshold value the output switches.

- They are very accurate, have high switching rate and can work in harsh environmental conditions.

- They have a limitation that they can detect only metallic targets and have limited operating range.

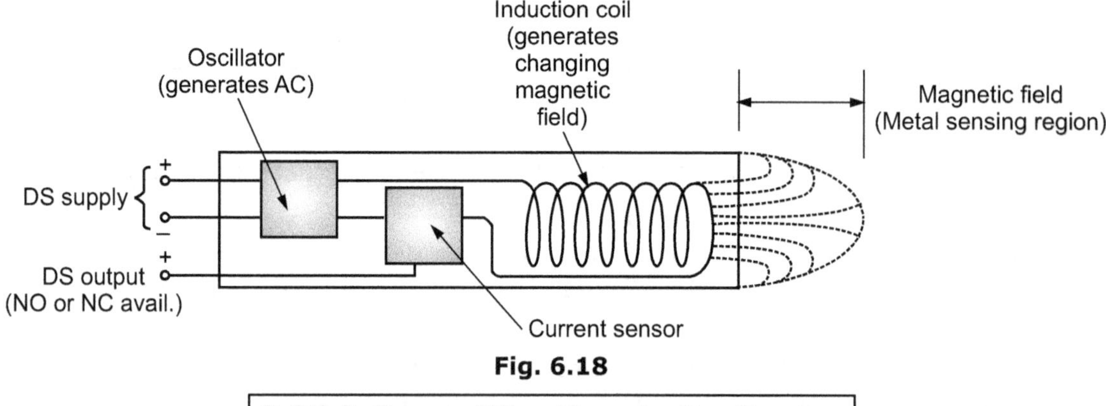

Fig. 6.18

6.22.3 Capacitive Proximity Sensor

- Capacitive proximity sensors are similar to inductive proximity sensors. The main difference between the two types is that capacitive proximity sensors produce an electrostatic field instead of an electromagnetic field.

- Capacitive proximity switches will sense metallic as well as non-metallic materials such as paper, wood, plastic, powder, glass, liquids and cloth without physical contact.

- The capacitive proximity sensor works on the capacitor principle. The main components of the capacitive proximity sensor are plate, oscillator, threshold detector and the output circuit.

- The plate inside the sensor acts as one plate of the capacitor and the target acts as another plate and the air acts as the dielectric between the plates.

- As the object comes close to the plate of the capacitor the capacitance increases, and as the object moves away from the plate, the capacitance decreases.

- The detector circuit checks the amplitude output from the oscillator and based on that the output switches.

- The capacitive sensor can detect any targets whose dielectric constant is more than air.

- It can detect both metallic and non-metallic targets, has good stability, speed, resolution and are of low cost.

- They have limitation that they are affected by temperature and humidity, could be triggered by dust, moisture, sensitive to noise, difficulties in designing.

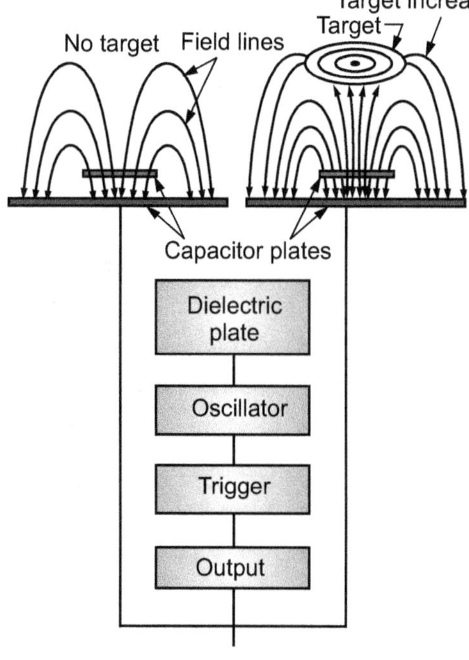

Fig. 6.19

6.22.4 Eddy Current Proximity Sensor

- When a coil is supplied with an alternating current input, an alternating magnetic field is produced around it. If there is a conducting material in close proximity to this alternating magnetic field, the eddy current is induced in it. This eddy current themselves produces magnetic field.

- An eddy current sensor typically has two coils. One of the coils generates a changing magnetic flux for reference. When this sensor comes close to any metal, an eddy current is induced in the metal, which in turn creates a magnetic flux opposite to the first coil's flux, effectively reducing the total flux. This change in the total flux is proportional to the proximity of that metal and is measured by the second coil.

- This result is send to the controller to take further action.
- These sensors are inexpensive, smaller in size, reliable and highly sensitive.
- They have a limitation that they can be used only for conductive materials only.

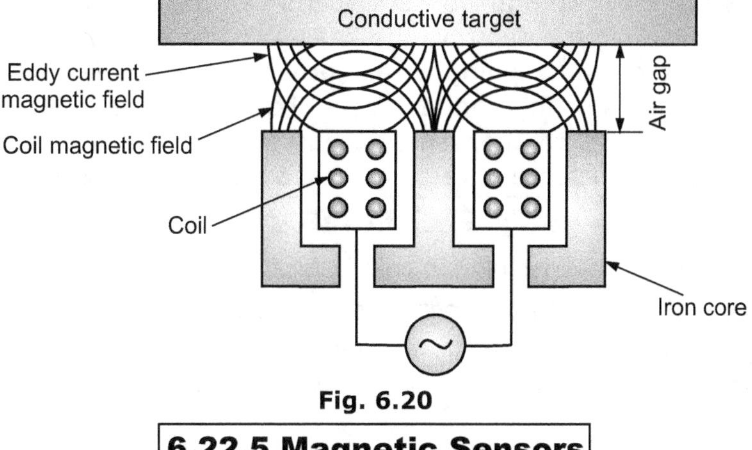

Fig. 6.20

6.22.5 Magnetic Sensors

- This technology operates by detecting changes in magnetic flux, induced by an applied force, through the use of Hall effect, magnetoresistive or magnetoelastic sensors.
- Hall effect sensors operate by measuring variations in the voltage that is generated by an electric current passing through a conductive material immersed in a magnetic field.
- Magnetoresistive and magnetoelastic sensors detect variations in magnetic fields generated by the application of mechanical stress.
- Applications are limited by the physical size of the sensing device, and by the need to operate in non-magnetic environments.

6.22.6 Hall Effect Sensors

- When a beam of charged particles passes through a magnetic field, the beam is deflected from its straight line path due to the forces acting on the particles. A current flowing in a conductor, such as a beam of moving charges and thus can be deflected by a magnetic field, called *Hall effect*.
- The working principle of a Hall Effect sensor is that 'a strip of conducting material carries a current in the presence of a transverse magnetic field' as shown in Fig. 6.21.
- The difference of potential is produced between the opposite edges of the conductor. The magnitude of the voltage depends upon the current and the magnetic field. The current is passed through leads 1 and 2 of the strip and the output leads 3 and 4 are connected with a Hall strip. When a transverse magnetic field passes through the strip, the voltage difference occurs in the output leads.

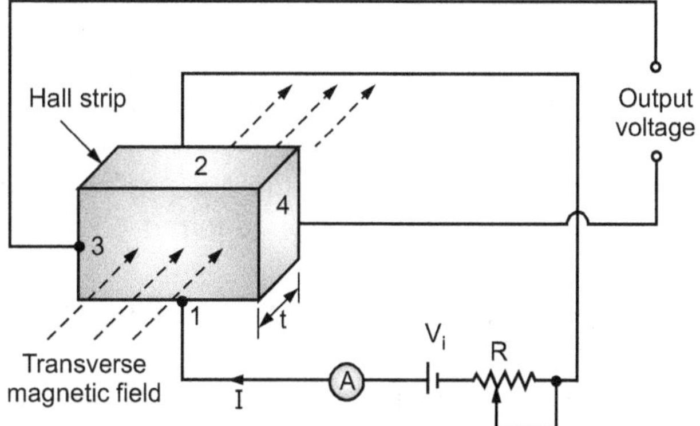

Fig. 6.21: Hall Effect Sensor

- Hall effect sensors are today available as an Integrated circuit with a necessary signal conditioning/processing circuitry.

6.22.7 Applications of Non-contact Sensors

- Counting the number products passed across a process. e.g. In an automotive industry, a non-contact sensor counts how many have left the painting area and how many have moved onto the curing area.
- Separating the defective jobs from jobs produced. e.g. if jobs produced are of different materials and colours, then they can be detected by means of colour sensor.

6.23 RANGE SENSORS

- The sensors those gather information without touching an object are called as *non*-Range sensors are devices that capture the 3D structure of the world from the viewpoint of the sensor, usually measuring the depth to the nearest surfaces. These measurements could be at a single point, across a scanning plane, or a full image with depth measurements at every point. The benefits of this range data is that a robot can be relatively certain where the real world is, relative to the sensor, thus allowing the robot to more reliably find navigable routes, avoid obstacles, grasp objects, act on industrial parts etc.

6.24 END EFFECTORS

- End effectors is an external mechanism that is located on the end of the manipulator.
- The end effector is the part of the robot that interacts with the environment.
- It can pick up objects, grasp them and manage their movement in accordance with the designed parameters. They are designed specifically to handle desired parts and to move in the intended path.
- An end effector is considered as the "hand of a robot". It is one of the important devices in a robot. It integrates an arm and a wrist, which helps it to perform several functions like material handling, pick and place, machine loading and unloading etc.
- It is also known as EOA (End of Arm Tooling), Robotic Peripherals and more.

6.25 FEATURES OF END EFFECTORS

- Any end effector should have the following essential features:
  1. It should provide a means of attaching the hand or tool to the robots wrist.
  2. It should provide a means to get the power for actuation of the tool.
  3. It should provide a means for proper mechanical linkage.
  4. It should provide a means for integrated sensor inside the tools.

6.26 CLASSIFICATION OF END EFFECTORS

- Based on the performance of work tasks, the robot end effector can be described into two important categories such as:
  - Grippers – to grasp and manipulate objects (e.g., parts) during work cycle.
  - Tools – to perform a process, e.g., spot welding, spray painting.

6.27 GRIPPERS

- Grippers are mainly used in a robot for grasping/holding an object. The grasped objects will be moved to the preferred place with the help of a robot.
- A gripper can be defined as 'a robot end effector that performs the operation to grasp temporarily, retain for required duration and release subsequently an object of a particular geometrical shape'.
- The gripper can be attached to the manipulator through a gripper plate as shown in Fig. 6.22 (a).
- The grippers are capable of carrying work parts, bottles, tools and so on. It can work by means of magnetic, mechanical, vacuum cups etc.
- It is often used to pick up dangerous or suspicious items for the robot to carry. Some can turn doorknobs and others are designed to carry only very specific things like beer cans.
- Closing too tightly on an object and crushing it is a major problem with autonomous grippers. There must be some way to tell how hard is enough to hold the object without dropping it or crushing it.
- Even for semi-autonomous robots where a human controls the manipulator, using the gripper effectively, is often difficult.

- For these reasons, gripper design requires as much knowledge as possible of the range of items the gripper will be expected to handle. Their mass, size, shape and strength etc. all must be taken into account.
- Some objects require grippers that have many jaws, but in most cases, grippers have only two jaws.
- There are several basic types of gripper geometries. A simple gripper with two jaws geared together is shown in Fig. 6.22 (b).
- Grippers can be classified into mechanical gripper, vacuum based, magnetic type, adhesive gripper etc.

6.27.1 Mechanical Gripper

- *The mechanical gripper* can be actuated by means of gear, cam, links, rope and pulley, screw etc.

6.27.2 Vacuum Gripper

- **Vacuum grippers** are vacuum actuated suction cups. The shape of cup depends upon the application for which it is to be used. Vacuum power is used to create grip and so used to lift smooth fragile objects such as plastic sheets, glass or thin metal sheets. But this consumes lot of power.
- **Vacuum End Effector** consists of a cup-shaped component and when it comes into contact with the object to be gripped, vacuum is created in the cup which ensures that the part remains attached. This type of end effector is used for delicate parts. The suction cup is connected by the mounting element to the robot arm. The unit consists of pneumatic or electric unit to generate or remove vacuum, valves and other elements to control the vacuum. A monitoring system ensures safe operation of the unit.

6.27.3 Magnetic Gripper

- *Magnetic End Effector* consists of a permanent magnet that moves in an aluminium cylinder. When the actuator drives the magnet towards the front end of the cylinder, it holds ferrous parts. As the magnet is extracted from the cylinder, the magnetic field fades and the parts are released. This type can be used for only ferrous parts and has the benefit of managing parts with asymmetrical form as well as holding a number of parts concurrently.

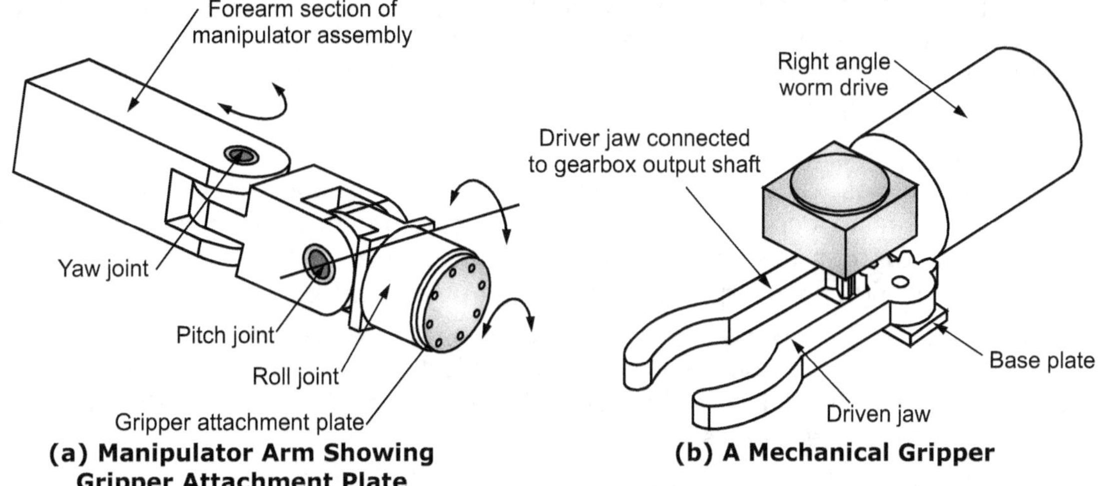

(a) Manipulator Arm Showing Gripper Attachment Plate   **(b) A Mechanical Gripper**

Fig. 6.22

6.28 ADVANTAGES OF ROBOT

- Industrial robots provides the following advantages:
  1. Robots can carry out work at high repeatability.
  2. Productivity can be improved as robots being machine do not need allowances.

3. They can work in unfavourable circumstances.
4. The manufacturing speed can be increased and the cost be reduced when robots are used.
5. Using robots may minimize the wasting of materials.
6. Using robots can improve the quality of a product by its ability to assemble precise components.
7. They can be used in assembly work with high precision and density.
8. The program can be modified easily, so robots can increase the flexibility in manufacturing.

6.29 RESTRICTIONS OF ROBOT

- Despite the advantages in using robots, there are a lot of restrictions as well.
    1. The cost for setting up a robot for industrial use is very high.
    2. It may sometimes be difficult to identify material because of limitations in the senses of robots.
    3. Robots cannot do very precise or delicate task like grinding of small gold jewellery, making pottery etc.
    4. The efficiency of robot also depends on the experience, knowledge and judgement of the programmer.
    5. The use of robot is not justified for production in small quantity.

6.30 ROBOT PROGRAMMING LANGUAGE

- Robots can perform arbitrary sequences of pre-stored motions or of motions computed as functions of sensory input. A robot can be applied to a large variety of tasks without significant redesign. This versatility comes from the overview of the robot's physical structure and control. But to take the benefit, the robot should be programmed easily.

- The lack of adequate programming tools can make some tasks impossible to perform. In other cases, the cost of programming may be a significant fraction of the total cost of an application. For these reasons, robot programming systems play a crucial role in robot development.

6.30.1 Approaches to Robot Programming

- There have been many styles of user interface developed for programming robots. Before the rapid proliferation of microcomputers in industry, robot controllers resembled the simple sequencers often used to control fixed automation. Modern approaches focus on computer programming, and issues in programming robots include all the issues faced in general computer programming.

Teach by Showing:
- The earliest and most widespread method of programming robots involves manually moving the robot to each desired position, and recording the internal joint co-ordinates corresponding to that position. In addition, operations such as closing the gripper or activating a welding gun are specified at some of these positions.

- The resulting "program" is a sequence of vectors of joint co-ordinates plus activation signals for external equipment. Such a program is executed by moving the robot through the specified sequence of joint co-ordinates and issuing the indicated signals. This method of programme is called, teach by showing.

(a) Explicit Robot Programming Languages:
- With the inception of powerful computers, the trend has been increasingly toward programming robots via programs written in computer programming languages. Usually, these computer programming languages have special features that apply to the problems of programming manipulators and so are called robot programming languages (RPLs).

- These robot programming languages have been built by developing a completely new language that, although addressing robot-specific areas, might well be considered a general computer programming language.
- An example is the VAL language developed to control the industrial robots. VAL was developed especially as a manipulator control language; as a general computer language, it was quite weak. For example, it did not support floating-point numbers or character strings, and subroutines could not pass arguments.

(b) Robot library for an existing computer language:

- These robot programming languages have been developed by starting with a popular computer language (e.g., Pascal) and adding a library of robot-specific subroutines. The user then writes a Pascal program making use of frequent calls to the predefined subroutine package for robot-specific needs.

(c) Robot library for a new general-purpose language:

- These robot programming languages have been developed by first creating a new general-purpose language as a programming base and then supplying a library of predefined robot-specific subroutines.
- Studies of actual application programs for robotic workcells have shown that a large percentage of the language statements are not robot-specific. Instead, a great deal of robot programming has to do with initialization, logic testing and branching, communication, and so on. For this reason, a trend might develop to move away from developing special languages for robot programming and move toward developing extensions to general languages, as in categories 2 and 3 above.

Task-level Programming Languages:

- The third level of robot programming methodology is task-level programming languages. In these languages the user does not have to specify the details of every action the robot has to take. User is allowed to command desired task directly. In such type of programming, the user can include instructions in the application program at a comparatively higher level. A task-level robot programming system must have the ability to perform many planning tasks automatically.
- For example, if an instruction to "grasp the welding rod" is issued, the system should be able to plan a path of the manipulator in such a way that there should not be any ambiguity with any surrounding obstacles. The robot must automatically choose a good grasp location on the rod and must grasp it.
- Whereas, in an explicit robot programming language, the programmer has to make all these choices. The border between explicit robot programming languages and task-level programming languages is quite distinct. Incremental advances are being made to explicit robot programming languages to help to ease programming, but these enhancements cannot be counted as components of a task-level programming system.

6.31 ROBOT DRIVE

- A robot requires a drive system to move its arm, wrist and body. A drive system is usually used to determine the capacity of a robot. For actuating the robot joints, there are three different types of drive systems available such as:
  - Electric drive system,
  - Hydraulic drive system, and
  - Pneumatic drive system.

6.31.1 Electric Drive System

- Electrical drive convert electrical signal commands into mechanical motions. The electric drive systems are capable of moving robots with higher speed or employ higher torque. The actuation of this type of robot can be done by either DC servo motors or DC stepper motors. It can be well suited for rotational joints and as well as linear joints.

- The electric drive system will be perfect for **small robots** and precise applications. Most importantly, it has got greater accuracy and repeatability. The one disadvantage of this system is that it is slightly costlier.

6.31.2 Hydraulic Drive System

- The hydraulic drive systems are completely meant for the **large – sized robots**. It can deliver high power or speed than the electric drive systems. This drive system can be used for both linear and rotational joints. The rotary motions are provided by the rotary vane actuators, while the linear motions are produced by hydraulic pistons. These systems can deliver a great deal of power compared to their actuator inertia.

- The robot using hydraulic drive can execute very quick movements with great force due to the low compressibility of hydraulic fluids and the high stiffness. It is reliable and relatively safe during operation. It has a low noise level.

- The **leakage** of hydraulic oils is considered as the major disadvantage of this drive. Any kind of leakage can also contaminate the oil and possibly lead to damage of interior surfaces.

- The hydraulic fluid is flammable and pressurized so leaks could pose an extreme hazard to equipment and personnel.

- There can be delays in the control of the system due to the transmission lines and oil viscosity changes from temperature.

6.31.3 Pneumatic Drive System

- The pneumatic drive systems are especially used for the **small type robots**, which have less than five degrees of freedom. It has the ability to offer fine accuracy and speed. This drive system can produce rotary movements by actuating the rotary actuators. The translational movements of sliding joints can also be provided by operating the piston. The price of this system is **less** when compared to the hydraulic drive.

- They are intrinsically safe and can be used in hazardous atmospheres. They are cheap compared to hydraulic systems (air costs nothing). A pneumatic actuator can generate more torque (force) to its own weight and thus have a better torque-weight ratio compared to an electrical actuator. The drawback of this system is that it will not be a perfect selection for the **faster operations.**

6.32 COMPARISON BETWEEN VARIOUS DRIVES

- Comparison between various drives is given below:

Parameter	Electric Drive	Pneumatic Drive	Hydraulic Drive
Source of energy	Electromagnetic field	Compressor	Electric motor
Input	Electric current	Compressed air (Dry fluid)	Pressurised oil (Liquid fluid)
Load application	Used for light load application	Used for medium load application	Used for heavy load applications
Storage of energy	Stores in batteries	Stores in accumulator	Stores in reservoir
Stroke length	Short movement via. solenoid	Medium movement via. pneumatic cylinder	Comparatively more movement via. hydraulic cylinder
Caution	Electric shocks	Air leakage and noise	Oil leakage and fire hazard
Applications	Motors used in toys, decoration	Pneumatic drill	In JCB cranes

6.33 APPLICATIONS OF ROBOT

- Robots are used in following applications:
  1. Robot can be successfully applied to the process of welding.
  2. Robot can perform processing operation like spray painting.
  3. In the process of assembly and inspection.
  4. Loading and unloading of machine components.
  5. Movement of material from one location to another.

6.34 MATERIAL HANDLING APPLICATION OF ROBOT

- Material handling applications are those in which the robot moves the materials or parts from one place to another. The robot is equipped with a gripper type of end-effector to accomplish this type of transfer. The gripper must be designed to handle the specific part or parts that are to be moved.

- In almost all the material handling applications, the part must be presented to the robot in familiar position and orientation.

- Within this application category, are the following cases:
  (i) Material transfer
  (ii) Machine loading/unloading.

Material Transfer:

- These are the operations in which the robot picks up the parts at one location and place them at a new location. The basic application in this category is pick and place operation, where robot picks up a part and deposits at a new location. Transferring parts from one conveyor to another is a classic example of this application.

- However, palletizing is a more complex example of the material transfer application. Here, the robots must retrieve parts, cartons, or other objects from one location and deposit them onto a pallet or other container with multiple locations.

Machine Loading/Unloading:

- In machine loading and unloading operations, the parts are transferred into/from a machine. The three possible scenarios can be machine loading, machine unloading, machine loading and unloading.

- In the machine loading operations, the robot loads parts into machine, but the parts are unloaded from the machine by some other mechanism. In the unloading operations, the machines are unloaded using the robots. When both the earlier situations are present, then this can be placed into the third category.

- Numerous applications of machine loading and unloading operations are as follows:
  (i)   Die casting operations
  (ii) Metal machining operations
  (iii) Plastic molding
  (iv) Forging
  (v)   Heat treating
  (vi) Press working robots as mentioned earlier are also used in the process industry.

- Numerous applications in this category are spot welding, continuous arc welding, spray painting, various rotating processes, and machining processes.

6.35 AUTOMATED GUIDED VEHICLES

- An automated guided vehicle is a material handling device that uses a driverless vehicle guided along defined pathways.

- The vehicle is powered by batteries that allow many hours of operation between recharging.

- A distinguish feature of an AGV is that, the pathways are unobstructed.

- An AGV is appropriate where different materials are moved from various load points to various unload points.
- The AGV can be divided into three types:
  1. Driverless trains
  2. Pallet trucks
  3. Unit load carriers

6.35.1 Driverless Trains

- A driverless train pulls one or more trailers to form a train as shown in Fig. 6.23.
- It was the first type of AGV to be introduced and is widely used today.
- For trains containing five to ten trailers, this is an efficient transport system.

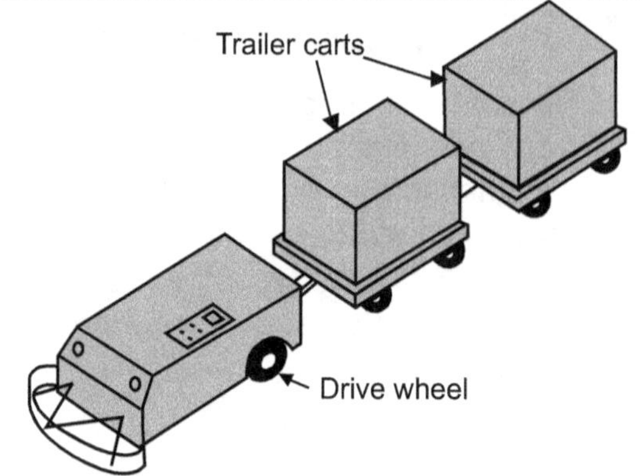

Fig. 6.23: Driverless Automated Guided Train

6.35.2 Pallet Trucks

- Automated guided pallet trucks are used to move palletized loads along predetermined routes.
- A pallet is loaded by a human worker as shown in Fig. 6.24, who steers the truck and uses its fork to elevate the load slightly then the worker drives the pallet truck to the guide path, programs its destination, and the vehicle proceeds automatically to the destination for unloading.

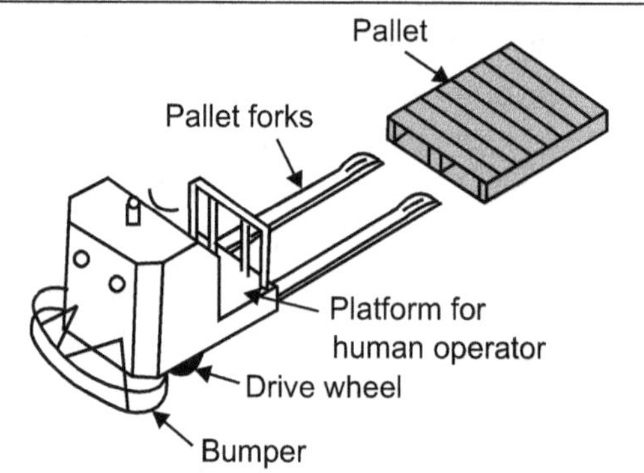

Fig. 6.24: Pallet Truck

6.35.3 Unit Load Carriers

- AGV unit load carriers are used to move unit loads from one station to another as shown in Fig. 6.25.
- They are often equipped for automatic loading and unloading of pallets by means of powered rollers, moving belts, mechanized lift platforms, or other devices built into the vehicle deck.
- The unit load carrier speed range around 50 m/min.

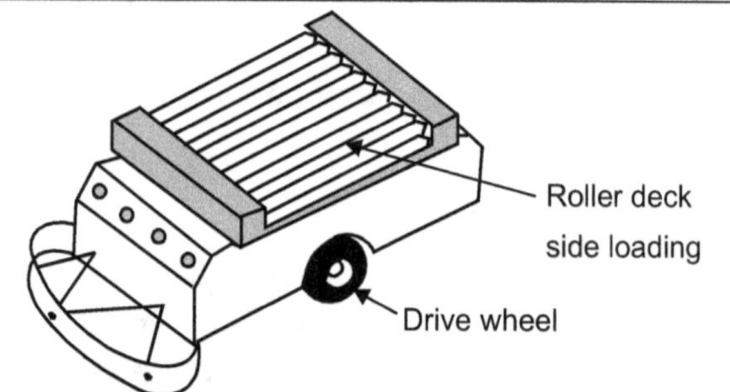

Fig. 6.25: Unit Load Carrier

6.36 VEHICLE GUIDANCE TECHNOLOGY

- The guidance system is a method by which AGV pathways are defined and vehicles are controlled to follow the pathways.
- The two methods used in commercial system for vehicle guidance are:
  (a) Imbedded Guide Wires.
  (b) Paint Strips.

6.36.1 Imbedded Guide Wires

- In the imbedded guide wire method, electrical wires are placed in a small channel cut into the surface of the floor.
- After the guide wire is installed, the channel is filled with cement to eliminate the discontinuity in the floor surface.
- The guide wire is connected to a frequency generator, which emits a low-voltage, low-current signal.
- This induces a magnetic field along the pathway that should be followed by sensors mounted on each vehicle.
- Two sensors are mounted on the vehicle on either side of the guide wire which does not allow the vehicle to leave its path.
- When a vehicle approaches a path which branches into two paths, the switch having different frequencies is used.
- As the vehicle enters into a branch, the switch identifies its path and determines its location.

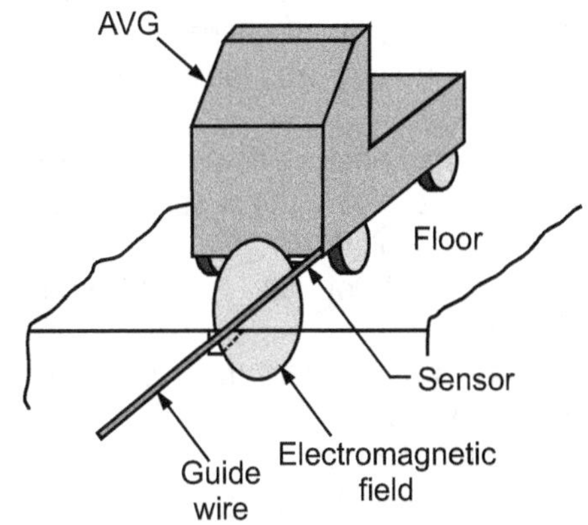

Fig. 6.26: Imbedded Guide Wires

6.36.2 Paint Strips

- When paint strips are used to define the pathway, the vehicle uses an optical sensor system capable of tracking the paint.
- The paint strips are painted or taped on the floor and contains fluorescent particles that reflect an UV light source from the vehicle.
- An on-board sensor detects the reflected light in the strip and controls the steering mechanism to follow it.
- Paint strips should be maintained clean and unscratched.

6.37 APPLICATIONS OF AGV

- Automated Guided Vehicles can be used in a wide variety of applications to transport many different types of material. The common uses are:
  1. Raw Material Handling.
  2. Work-in-Process Movement.
  3. Pallet Handling.
  4. Finished Product Handling.
  5. Trailer Loading.
  6. Paper Roll Handling.
  7. Container Handling.

6.38 ADVANTAGES OF AGV

- AGV has many advantages. Some of them are given below:
  1. Dispatching, tracking and monitoring under real time control.
  2. Better resource utilisation (justify in three years).
  3. Increased control over material flow and movement.
  4. Reduced material damage and less material movement noise.
  5. Routing consistency and flexibility.
  6. Operational reliability in hazardous and special environment.
  7. Ability to interface with machine tool, robots, conveyors etc.

Important Points

Definition of Robot and Robotics:

- Robotics Institute of America defines robot as, "A robot is a reprogrammable multifunctional manipulator designed to move material, parts, tools or specialized devices through variable programmed motions for the performance of a variety of tasks".
- A mechanical device with joints and links, guided by sensors and driven by actuators, controlled by a program, that handle and manipulate parts, materials, tools and devices for performing different tasks in various work conditions can be called as a robot.
- Robotics is 'a branch of engineering that involves the conception, design, manufacture and operation of robots'. This field overlaps with electronics, computer science, artificial intelligence, mechatronics, nanotechnology and bioengineering.

Laws of Robotics:

- A robot may not injure humanity, or through inaction, allow humanity to come to harm.
- A robot may not injure a human being, or, through inaction, allow a human being to come to harm, unless this would violate a higher order law.
- A robot must obey the orders given to it by human beings except where such orders would conflict with the First Law.
- A robot must protect its own existence as long as such protection does not conflict with the First or Second Law.

Characteristics of Robot:

- **Sensing:** Robot should have the ability to sense its surroundings. This needs a lot of sensors to be incorporated in the robot.
- **Movement:** A robot needs to be able to move around its surrounding.
- **Energy:** A robot needs to be able to power itself. A robot might be solar powered, electrically powered, battery powered.
- **Intelligence:** Robot can use logic by means of programming. A program is stored in the robot controller which can be called as the brain of robot.

Degree of Freedom:

- It refers to the ability of the robot arm to move forward and backward, up and down and to the left and to the right.
- For each degree of freedom, a joint is required. A robot requires six degrees of freedom to be completely versatile.
- The three basic degrees of freedom in the robot arm are one rotational and two linear.
- For applications that require more freedom, additional degrees can be obtained from the wrist, which can be obtained from the end effector.
- These degree of freedom are named as: pitch, yaw and roll.

Work Envelope of Robot:

- Work envelope is also known as work space or work volume. It is the region or space the end effector of the manipulator can reach.
- The work envelope is the range of motion over which a robot arm can move.
- When choosing a robot arm for a certain industrial purpose, it is important that the work envelope be large enough to encompass all the points that the robot arm will be required to reach.

Basic Components of Robot:

- **Arm:** The arm is the part of the robot that positions the end-effector and sensors to do their pre-programmed task.
- **End-effectors:** A hand of a robot is considered as end effector. The effectors are the parts of the robot that actually do the work.
- **Actuators:** An actuator is a device that produces translatory or rotary movement in the links or makes the freedoms possible.
- **Sensors:** A sensor is an element in a control system that acquires a physical parameter and changes it into a signal.
- **Controllers:** Every robot is connected to a computer that co-ordinates both the arm to work together. This computer is known as the controller.
- **Drive:** The drive is the engine or motor that moves the links into their designated positions.

Robot Anatomy:
- The Anatomy of Industrial Robots deals with the assembling of outer components of a robot such as wrist, arm and body.
- Robot manipulator consists of Body-and-arm assembly and Wrist assembly.
- At the end of the manipulator's wrist assembly is a device related to the task that must be accomplished by the robot. The device is called an end effector and can be either a gripper for holding a work part, or a specific tool for performing some process.

Joints and Links:
- **Linear joint:** The linear joint provides the translational sliding motion between the input and output link.
- **Orthogonal joint:** The orthogonal joint provides the translational sliding motion between the input and output link.
- **Rotational joint:** The rotational joint provides the relative rotational motion between the input and output link.
- **Twisting joint:** The twisting joint provides the relative rotational motion between the input and output link.
- **Revolving joint:** The revolving point provides the same as twisting joint.

Basic Robot Configuration:

Polar Configuration:
- In the polar configuration, the robot arm has following movements:
  - Linear movement that allows the arm to extend and retract because of one linear joint.
  - Rotary movement that occurs around an axis (vertical) perpendicular to the base because of one twisting joint.
  - Vertical lift of the arm about the pivot point because of one rotational joint.

Cylindrical Configuration:
- In the cylindrical configuration, the robot arm has following movements:
  - Rotational movement of the column about its axis because of one twisting joint.
  - Linear movement of the assembly along the column because of one linear joint.
  - Linear movement in and out relative to the column axis because of one orthogonal joint.

Cartesian Co-ordinate Configuration:
- In the Cartesian co-ordinate configuration, the robot arm has following movements:
  - Linear movement that allows vertical lift to the arm because of one linear joint.
  - Two sliding movements those are perpendicular to each other because of two orthogonal joints.

Jointed arm Configuration:
- In this configuration, the robot arm has following movements:
  - Rotary movement (vertical column that swivels about the base) that occurs around an axis (horizontal) parallel to the base because of one twisting joint.
  - Rotary movement at the top of the column about the shoulder joint (along horizontal axis) because of one rotational joint.
  - Rotary movement at the output arm about the elbow joint (along horizontal axis) because of one rotational joint.

Scara:
- In Scara, the robot arm has following movements:
  - Linear movement that allows the arm to extend and retract because of one orthogonal joint.
  - Rotary movement at the top of the column about the shoulder joint (along vertical axis) because of one revolving joint.
  - Rotary movement at the output arm about the elbow joint (along vertical axis) because of one rotational joint.

Robot Sensors:

Contacting Type Sensors:
- The sensors that operate on touch are called as contact sensors. As these sensors must touch their environment to operate, their use is limited to objects and conditions that can do no harm to the sensors.

Touch Sensors:
- It simply indicate whether physical contact is been made with the object.
- The touch sensors are basically used to determine whether the gripper jaw touches any object. This may also send a signal to the controller that target is reached, collision may occur.

Force Sensors:
- It is used to indicate the magnitude of force with the object.
- These sensors are located between the gripper and the last joint of the wrist so that the forces can be measured.

Non-contacting Type Sensors:
- The sensors those gather information without touching an object are called as non-contact sensors.
- They can be used in environments where contact sensors would get damaged or wear out quickly.

Optical Type Sensors:
- Photocells and other photometric devices that are used to detect the presence or absence of objects. Often used in conjunction with proximity sensors.

Proximity Sensors:
- These sensors are used to determine that an object is close to the sensor before it makes contact. These are non-contact type of devices that indicate the presence of an object within specific distance from the sensor.

Range Sensors:
- Sensors that gather information without touching an object.
- Devices that capture the 3D structure of the world from the viewpoint of the sensor.
- Benefit is that, a robot can be relatively certain where the real world is.

End Effectors:
- End effectors are an external mechanism that is located on the end of the manipulator.
- It can pick up objects, grasp them and manage their movement in accordance with the designed parameters. They are designed specifically to handle desired parts and to move in the intended path.

Features of End Effectors:
- It should provide a means of attaching the hand or tool to the robots wrist.
- It should provide a means to get the power for actuation of the tool.
- It should provide a means for proper mechanical linkage.
- It should provide a means for integrated sensor inside the tools.

Types of End Effectors:
Grippers:
- The grippers are mainly used in a robot for grasping/holding an object. The grasped objects will be moved to the preferred place with the help of a robot.
- The gripper can be attached to the manipulator through a gripper plate.

Advantages of Robot:
- Robots can carry out work at high repeatability.
- Productivity can be improved as robots being machine do not need allowances.
- They can work in unfavourable circumstances.
- The manufacturing speed can be increased and the cost be reduced when robots are used.
- Using robots may minimize the wasting of materials.
- Using robots can improve the quality of a product by its ability to assemble precise components.
- They can be used in assembly work with high precision and density.
- The program can be modified easily so robots can increase the flexibility in manufacturing.

Restrictions of Robot:
- The cost for setting up a robot for industrial use is very high.
- It may sometimes be difficult to identify material because of limitations in the senses of robots.
- Robots cannot do very precise or delicate task like grinding of small gold jewellery, making pottery etc.
- The efficiency of robot also depends on the experience, knowledge and judgement of the programmer.
- The use of robot is not justified for production in small quantity.

Applications of Robot:
- Robot can be successfully applied to the process of welding.
- Robot can perform processing operation like spray painting.
- In the process of assembly and inspection.
- Loading and unloading of machine components.
- Movement of material from one location to another.

Robot Programming Language:
- Teach by Showing
- Task-level Programming Languages

Robot Drive:
- There are three different types of drive systems available such as:
  - Electric drive system,
  - Hydraulic drive system, and
  - Pneumatic drive system.

Material Handling Application of Robot:
- Within this application category, are the following cases:
  - (i) Material transfer
  - (ii) Machine loading/unloading.

Automated Guided Vehicle:
- An automated guided vehicle is a material handling device that uses a driverless vehicle guided along defined pathways.
- The vehicle is powered by batteries that allow many hours of operation between recharging.
- An AGV is appropriate where different materials are moved from various load points to various unload points.

Practice Questions

1. Define robot and robotics.
2. State the laws of robotics.
3. State the characteristics of robot.
4. Explain degree of freedom in robots.
5. Explain the concept of work envelop.
6. Explain the basic components of robots.
7. Describe the working principle of robot.
8. Explain the joints and links with neat sketch.
9. Explain various robot configurations with sketches.
10. Give functions of robot sensor.
11. Explain contact sensors.
12. Explain non-contact sensors.
13. Explain proximity sensors.
14. Describe range sensors.
15. State the features of end effector.
16. Give the classification of end effector.
17. Explain robot grippers.
18. Give advantages of robot.
19. State restrictions for robots.
20. Explain robot programming language.
21. Describe various drives in robot.
22. Compare various robot drives.
23. Give applications of robots.
24. Explain material handling application of robot.
25. Explain automated guided vehicle.

■ ■ ■